T0331880

Mathematical Modeling in Chemical Engineering

A solid introduction to mathematical modeling for a range of chemical engineering applications, covering model formulation, simplification, and validation. It explains how to describe a physical/chemical reality in mathematical language and how to select the type and degree of sophistication for a model. Model reduction and approximate methods are presented, including dimensional analysis, time constant analysis, and asymptotic methods. An overview of solution methods for typical classes of models is given. As final steps in model building, parameter estimation and model validation, and assessment are discussed. The reader is given hands-on experience of formulating new models, reducing the models, and validating the models.

The authors assume a knowledge of basic chemical engineering, in particular transport phenomena, as well as basic mathematics, statistics, and programming. The accompanying problems, tutorials, and projects include model formulation at different levels, analysis, parameter estimation, and numerical solution.

Anders Rasmuson is a Professor in Chemical Engineering at Chalmers University of Technology, Gothenburg, Sweden. He obtained his Ph.D. in chemical engineering at the Royal Institute of Technology, Stockholm, in 1978. His research has focused on mathematical modeling combined with experimental work in many areas of chemical engineering, for example particulate processes, multiphase flows and transport phenomena, mixing, and separation processes.

Bengt Andersson is a Professor in Chemical Engineering at Chalmers University of Technology. He obtained his Ph.D. in chemical engineering at Chalmers in 1977. His research has focused on experimental studies and modeling of mass and heat transfer in various chemical reactors, ranging from automotive catalysis to three-phase flow in chemical reactors.

Louise Olsson is a Professor in Chemical Engineering at Chalmers University of Technology. She obtained her Ph.D. in chemical engineering at Chalmers in 2002. Her research has focused on experiments and kinetic modeling of heterogeneous catalysis.

Ronnie Andersson is an Assistant Professor in Chemical Engineering at Chalmers University of Technology. He obtained his Ph.D. at Chalmers in 2005, and from 2005 until 2010 he worked as consultant at Epsilon HighTech as a specialist in computational fluid dynamic simulations of combustion and multiphase flows. His research projects involve physical modeling, fluid dynamic simulations, and experimental methods.

Mathematical Modeling in Chemical Engineering

ANDERS RASMUSON

Chalmers University of Technology, Gothenberg

BENGT ANDERSSON

Chalmers University of Technology, Gothenberg

LOUISE OLSSON

Chalmers University of Technology, Gothenberg

RONNIE ANDERSSON

Chalmers University of Technology, Gothenberg

CAMBRIDGE
UNIVERSITY PRESS

CAMBRIDGE
UNIVERSITY PRESS

University Printing House, Cambridge CB2 8BS, United Kingdom

One Liberty Plaza, 20th Floor, New York, NY 10006, USA

477 Williamstown Road, Port Melbourne, VIC 3207, Australia

4843/24, 2nd Floor, Ansari Road, Daryaganj, Delhi - 110002, India

79 Anson Road, #06-04/06, Singapore 079906

Cambridge University Press is part of the University of Cambridge.

It furthers the University's mission by disseminating knowledge in the pursuit of education, learning and research at the highest international levels of excellence.

www.cambridge.org
Information on this title: www.cambridge.org/9781107049697

© Cambridge University Press 2014

First published 2014

A catalogue record for this publication is available from the British Library

Library of Congress Cataloging in Publication data
Rasmuson, Anders, 1951–
Mathematical modeling in chemical engineering / Anders Rasmuson, Chalmers
University of Technology, Gothenberg, Bengt Andersson, Chalmers University of
Technology, Gothenberg, Louise Olsson, Chalmers University of Technology,
Gothenberg, Ronnie Andersson, Chalmers University of Technology, Gothenberg.
    pages   cm
Includes index.
ISBN 978-1-107-04969-7 (hardback)
1. Chemical engineering – Mathematical models.   I. Title.
TP155.2.M35R37   2014
660 – dc23   2013040670

ISBN 978-1-107-04969-7 Hardback

Contents

Preface   *page* ix

1   Introduction   1

   1.1   Why do mathematical modeling?   1
   1.2   The modeling procedure   5
   1.3   Questions   9

2   Classification   10

   2.1   Grouping of models into opposite pairs   10
   2.2   Classification based on mathematical complexity   14
   2.3   Classification according to scale (degree of physical detail)   16
   2.4   Questions   18

3   Model formulation   20

   3.1   Balances and conservation principles   20
   3.2   Transport phenomena models   22
   3.3   Boundary conditions   26
   3.4   Population balance models   28
       3.4.1   Application to RTDs   32
   3.5   Questions   34
   3.6   Practice problems   35

4   Empirical model building   40

   4.1   Dimensional systems   40
   4.2   Dimensionless equations   41
   4.3   Empirical models   46
   4.4   Scaling up   48
   4.5   Practice problems   52

5   Strategies for simplifying mathematical models   53

   5.1   Reducing mathematical models   54

	5.1.1	Decoupling equations	55
	5.1.2	Reducing the number of independent variables	55
	5.1.3	Lumping	56
	5.1.4	Simplified geometry	56
	5.1.5	Steady state or transient	58
	5.1.6	Linearizing	61
	5.1.7	Limiting cases	63
	5.1.8	Neglecting terms	64
	5.1.9	Changing the boundary conditions	67
5.2	Case study: Modeling flow, heat, and reaction in a tubular reactor		68
	5.2.1	General equation for a cylindrical reactor	68
	5.2.2	Reducing the number of independent variables	69
	5.2.3	Steady state or transient?	70
	5.2.4	Decoupling equations	72
	5.2.5	Simplified geometry	72
	5.2.6	Limiting cases	75
	5.2.7	Conclusions	76
5.3	Error estimations		76
	5.3.1	Sensitivity analysis	77
	5.3.2	Over- and underestimations	77
5.4	Questions		77
5.5	Practice problems		78

6   Numerical methods   81

6.1	Ordinary differential equations		81
	6.1.1	ODE classification	81
	6.1.2	Solving initial-value problems	82
	6.1.3	Numerical accuracy	87
	6.1.4	Adaptive step size methods and error control	88
	6.1.5	Implicit methods and stability	90
	6.1.6	Multistep methods and predictor–corrector pairs	93
	6.1.7	Systems of ODEs	94
	6.1.8	Transforming higher-order ODEs	96
	6.1.9	Stiffness of ODEs	97
6.2	Boundary-value problems		99
	6.2.1	Shooting method	99
	6.2.2	Finite difference method for BVPs	102
	6.2.3	Collocation and finite element methods	107
6.3	Partial differential equations		108
	6.3.1	Classification of PDEs	109
	6.3.2	Finite difference solution of parabolic equations	110
	6.3.3	Forward difference method	110
	6.3.4	Backward difference method	113

6.3.5   Crank–Nicolson method 114
6.4   Simulation software 114
6.4.1   MATLAB 114
6.4.2   Miscellaneous MATLAB algorithms 115
6.4.3   An example of MATLAB code 116
6.4.4   GNU Octave 117
6.5   Summary 117
6.6   Questions 117
6.7   Practice problems 118

7   Statistical analysis of mathematical models 121

7.1   Introduction 121
7.2   Linear regression 121
7.2.1   Least squares method 123
7.3   Linear regression in its generalized form 125
7.3.1   Least square method 126
7.4   Weighted least squares 127
7.4.1   Stabilization of the variance 127
7.4.2   Placing greater/less weight on certain experimental parts 128
7.5   Confidence intervals and regions 130
7.5.1   Confidence intervals 130
7.5.2   Student's t-tests of individual parameters 133
7.5.3   Confidence regions and bands 134
7.6   Correlation between parameters 136
7.6.1   Variance and co-variance 136
7.6.2   Correlation matrix 137
7.7   Non-linear regression 138
7.7.1   Intrinsically linear models 138
7.7.2   Non-linear models 139
7.7.3   Approximate confidence levels and regions for non-linear models 140
7.7.4   Correlation between parameters for non-linear models 142
7.8   Model assessments 142
7.8.1   Residual plots 142
7.8.2   Analysis of variance (ANOVA) table 145
7.8.3 R^2 statistic 149
7.9   Case study 7.1: Statistical analysis of a linear model 149
7.9.1   Solution 150
7.10   Case study 7.2: Multiple regression 153
7.10.1   Solution 154
7.11   Case study 7.3: Non-linear model with one predictor 158
7.11.1   Solution 159
7.12   Questions 163
7.13   Practice problems 163

Appendix A Microscopic transport equations   168
Appendix B Dimensionless variables   170
Appendix C Student's t-distribution   173

Bibliography   180
Index   181

Preface

The aim of this textbook is to give the reader insight and skill in the formulation, construction, simplification, evaluation/interpretation, and use of mathematical models in chemical engineering. It is *not* a book about the solution of mathematical models, even though an overview of solution methods for typical classes of models is given.

Models of different types and complexities find more and more use in chemical engineering, e.g. for the design, scale-up/down, optimization, and operation of reactors, separators, and heat exchangers. Mathematical models are also used in the planning and evaluation of experiments and for developing mechanistic understanding of complex systems. Examples include balance models in differential or integral form, and algebraic models, such as equilibrium models.

The book includes model formulation, i.e. how to describe a physical/chemical reality in mathematical language, and how to choose the type and degree of sophistication of a model. It is emphasized that this is an iterative procedure where models are gradually refined or rejected in confrontation with experiments. Model reduction and approximate methods, such as dimensional analysis, time constant analysis, and asymptotic methods, are treated. An overview of solution methods for typical classes of models is given. Parameter estimation and model validation and assessment, as final steps, in model building are discussed. The question "What model should be used for a given situation?" is answered.

The book is accompanied by problems, tutorials, and projects. The projects include model formulation at different levels, analysis, parameter estimation, and numerical solution.

The book is aimed at chemical engineering students, and a basic knowledge of chemical engineering, in particular transport phenomena, will be assumed. Basic mathematics, statistics, and programming skills are also required.

Using the book (course) the reader should be able to construct, solve, and apply mathematical models for chemical engineering problems. In particular:

- construct models using balances on differential or macroscopic control volumes for momentum, heat, mass, and numbers (population balances);
- construct models by simplification of general model equations;

- understand and use methods for model simplification;
- understand differences between models;
- understand and use numerical solution methods;
- understand and perform parameter estimation;
- use model assessment techniques to be able to judge if a model is good enough.

1 Introduction

In this introductory chapter the use of mathematical models in chemical engineering is motivated and examples are given. The general modeling procedure is described, and some important tools that are covered in greater detail later in the book are outlined.

1.1 Why do mathematical modeling?

Mathematical modeling has always been an important activity in science and engineering. The formulation of qualitative questions about an observed phenomenon as mathematical problems was the motivation for and an integral part of the development of mathematics from the very beginning.

Although problem solving has been practiced for a very long time, the use of mathematics as a very effective tool in problem solving has gained prominence in the last 50 years, mainly due to rapid developments in computing. Computational power is particularly important in modeling chemical engineering systems, as the physical and chemical laws governing these processes are complex. Besides heat, mass, and momentum transfer, these processes may also include chemical reactions, reaction heat, adsorption, desorption, phase transition, multiphase flow, etc. This makes modeling challenging but also necessary to understand complex interactions.

All models are abstractions of real systems and processes. Nevertheless, they serve as tools for engineers and scientists to develop an understanding of important systems and processes using mathematical equations. In a chemical engineering context, mathematical modeling is a prerequisite for:

- design and scale-up;
- process control;
- optimization;
- mechanistic understanding;
- evaluation/planning of experiments;
- trouble shooting and diagnostics;
- determining quantities that cannot be measured directly;
- simulation instead of costly experiments in the development lab;
- feasibility studies to determine potential before building prototype equipment or devices.

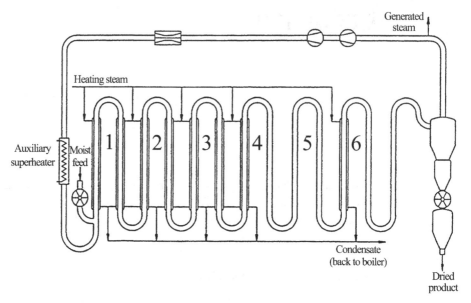

Figure 1.1. Pilot dryer, Example 1.1.

A typical problem in chemical engineering concerns scale-up from laboratory to full-scale equipment. To be able to scale-up with some certainty, the fundamental mechanisms have to be evaluated and formulated in mathematical terms. This involves careful experimental work in close connection to the theoretical development.

There are no modeling recipes that guarantee successful results. However, the development of new models always requires both an understanding of the physical/chemical principles controlling a process and the skills for making appropriate simplifying assumptions. Models will never be anything other than simplified representations of real processes, but as long as the essential mechanisms are included the model predictions can be accurate. Chapter 3 therefore provides information on how to formulate mathematical models correctly and Chapter 5 teaches the reader how to simplify the models.

Let us now look at two examples and discuss the mechanisms that control these systems. We do this without going into the details of the formulation or numerical solution. After reading this book, the reader is encouraged to refer back to these two case studies and read how these modeling problems were solved.

Example 1.1 Design of a pneumatic conveying dryer

A mathematical model of a pneumatic conveying dryer, Figure 1.1, has been developed (Fyhr, C. and Rasmuson, A., *AIChE J.* **42**, 2491–2502, 1996; **43**, 2889–2902, 1997) and validated against experimental results in a pilot dryer.

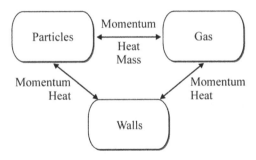

Figure 1.2. Interactions between particles, steam, and walls.

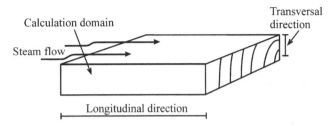

Figure 1.3. Wood chip.

The dryer essentially consists of a long tube in which the material is conveyed by, in our case, superheated steam. The aim of the modeling task was to develop a tool that could be used for design and rating purposes.

Inside the tubes, the single particles, conveying steam, and walls interact in a complex manner, as illustrated in Figure 1.2. The gas and particles exchange heat and mass due to drying, and momentum in order to convey the particles. The gas and walls exchange momentum by wall friction, as well as heat by convection. The single particles and walls also exchange momentum by wall friction, and heat by radiation from the walls. The single particle is, in this case, a wood chip shaped as depicted in Figure 1.3.

The chip is rectangular, which leads to problems in determining exchange coefficients. The particles also flow in a disordered manner through the dryer. The drying rate is controlled by external heat transfer as long as the surface is kept wet. As the surface dries out, the drying rate decreases and becomes a function of both the external and internal characteristics of the drying medium and single particle. The insertion of cold material into the dryer leads to the condensation of steam on the wood chip surface, which, initially, increases the moisture content of the wood chip. The pressure drop at the outlet leads to flashing, which, in contrast, reduces the moisture content.

The mechanisms that occur between the particles and the steam, as well as the mechanisms inside the wood chip, are thus complex, and a detailed understanding is necessary. How would you go about modeling this problem? Models for these complex processes have been developed in the cited articles by Fyhr and Rasmuson.

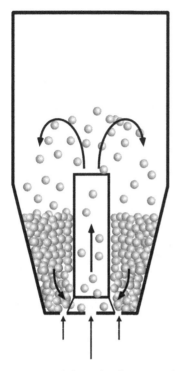

Figure 1.4. Schematic of a Wurster bed, Example 1.2.

Example 1.2 Design and optimization of a Wurster bed coater

In the second example, a mathematical model of a Wurster bed coater, Figure 1.4, has been developed (Karlsson, S., Rasmuson, A., van Wachen, B., and Niklasson Bjorn, I., *AIChE J.* **55**, 2578–2590, 2009; Karlsson, S., Rasmuson, A., Niklasson Bjorn, I., and Schantz, S., *Powder Tech.* **207**, 245–256, 2011) and validated against experimental results.

Coating is a common process step in the chemical, agricultural, pharmaceutical, and food industries. Coating of solid particles is used for the sustained release of active components, for protection of the core from external conditions, for masking taste or odours, and for easier powder handling. For example, several applications in particular are used for coating in the pharmaceutical industry, for both aesthetic and functional purposes.

The Wurster process is a type of spouted bed with a draft tube and fluidization flow around the jet (Figure 1.4). The jet consists of a spray nozzle that injects air and droplets of the coating liquid into the bed. The droplets hit and wet the particles concurrently in the inlet to the draft tube. The particles are transported upwards through the tube, decelerate in the expansion chamber, and fall down to the dense region of particles outside the tube. During the upward movement and the deceleration, the particles are dried by the warm air, and a thin coating layer starts to form on the particle surface. From the dense region the particles are transported again into the Wurster tube, where the droplets again hit the particles, and the circulation motion in the bed is repeated. The particles are circulated until a sufficiently thick layer of coating material has been built up around them.

The final coating properties, such as film thickness distribution, depend not only on the coating material, but also on the process equipment and the operating conditions during film formation. The spray rate, temperature, and moisture content are operating parameters that influence the final coating and which can be controlled in the process. The drying rate and the subsequent film formation are highly dependent on the flow field of the gas and the particles in the equipment. Local temperatures in the equipment are also known to be critical for the film formation; different temperatures may change the properties of the coating layer. Temperature is also important for moisture equilibrium, and influences the drying rate.

Several processes take place simultaneously at the single-particle level during the coating phase. These are: the atomization of the coating solution, transport of the droplets formed to the particle, adhesion of the droplets to the particle surface, surface wetting, and film formation and drying. These processes are repeated for each applied film layer, i.e. continuously repeated for each circulation through the Wurster bed.

Consequently, the mechanisms that occur at the microscopic and macroscopic levels are complex and include a high degree of interaction. The aim of the modeling task is to develop a tool that can be used for design and optimization. What models do you think best describe the mechanisms in this process?

1.2   The modeling procedure

In undergraduate textbooks, models are often presented in their final, neat and elegant form. In reality there are many steps, choices, and iterative processes that a modeler goes through in reaching a satisfactory model. Each step in the modeling process requires an understanding of a variety of concepts and techniques blended with a combination of critical and creative thinking, intuition and foresight, and decision making. This makes model building both a science and an art.

Model building comprises different steps, as shown in Figure 1.5. As seen here, model develpoment is an iterative process of hypotheses formulation, validation, and refinement.

Figure 1.5 also gives an outline of this process. Conceptual and mathematical model formulation are treated further in Chapters 3–5; solution methods are discussed in Chapter 6; and finally parameter estimation and model validation are discussed in Chapter 7.

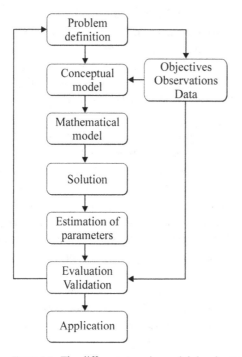

Figure 1.5. The different steps in model development.

Step 1: Problem definition
The first step in the mathematical model development is to define the problem. This involves stating clear goals for the modeling, including the various elements that pertain to the problem and its solution.

Consider the following questions:

What is the objective (i.e. what questions should the model be able to answer)?
What resolution is needed?
What degree of accuracy is required?

Step 2: Formulation of conceptual model (Chapter 3)
When formulating the conceptual model, decisions must be made on what hypothesis and which assumptions to use. The first task is to collect data and experience about the subject to be modeled. The main challenges are in identifying the underlying mechanisms and governing physical/chemical principles of the problem. The development of a conceptual model involves idealization, and there will always be a tradeoff between model generality and precision.

Step 3: Formulation of mathematical model (Chapters 3–5)
Each important quantity is represented by a suitable mathematical entity, e.g. a variable, a function, a graph, etc.

What are the variables (dependent, independent, parameters)? The distinction between dependent and independent variables is that the independent variable is the one being changed, x, and the dependent variable, y, is the observed variable caused by this change, e.g. $y = x^3$. Parameters represent physical quantities that characterize the system and model such as density, thermal conductivity, viscosity, reaction rate constants, or activation energies. Parameters are not necessarily constants, and can be described as functions of the dependent (or independent) variables, e.g. heat capacity $c_p(T)$ and density $\rho(p, T)$.

What are the constraints? Are there limitations on the possible values of a variable? For example, concentrations are always positive.

What boundary conditions, i.e. the relations valid at the boundaries of the system, are suitable to use?

What initial conditions, i.e. conditions valid at the start-up of a time-dependent process, exist?

Each relationship is represented by an equation, inequality, or other suitable mathematical relation.

Step 4: Solution of the mathematical problem (Chapter 6)
Check the validity of individual mathematical relationships, and whether the relationships are mutually consistent.

Consider the analytical versus the numerical solution. Analytical solutions are only possible for special situations; essentially the problem has to be linear. Most often, a numerical solution is the only option; luckily the cost of computers is low and models can run in parallel on computer clusters if necessary.

Verify the mathematical solution, i.e. ensure that you have solved the equations correctly. This step involves checking your solution against previously known results (analytical/numerical), simplified limiting cases, etc.

Step 5: Estimation of parameters (Chapter 7)
The parameters of the system must be evaluated and the appropriate values must be used in the model. Some parameters can be obtained independently of the mathematical model. They may be of a basic character, like the gravitation constant, or it may be possible to determine them by independent measurements, like, for instance, solubility data from solubility experiments. However, it is usually not possible to evaluate all the parameters from specific experiments, and many of them have to be estimated by taking results from the whole (or a similar system), and then using parameter-fitting techniques to determine which set of parameter values makes the model best fit the experimental results. For example, a complex reaction may involve ten or more kinetic constants. These constants can be estimated by fitting a model to results from a laboratory reactor. Once the parameter values have been determined, they can be incorporated into a model of a plant-scale reactor.

Step 6: Evaluation/validation (Chapter 7)
A key step in mathematical modeling is experimental validation. Ideally the validation should be made using independent experimental results, i.e. not the same set as used

for parameter estimation. During the validation procedure it may happen that the model still has some deficiencies. In that case, we have to "iterate" the model and eventually modify it. In the work by Melander, O. and Rasmuson, A. (*Nordic Pulp Paper Res. J.* **20**, 78–86, 2005) it was found that the original model for pulp fiber flow in a gas stream severely underestimated lateral spreading of the fibers. Detailed analysis led to a modified model (Melander, O. and Rasmuson, A., *J. Multiphase Flow* **33**, 333–346, 2007) with an additional term in the governing equations, and good agreement with experimental data.

In Chapter 7, the general question of model quality is discussed. Is the model good enough?

In the evaluation of the model, sensitivity analysis, i.e. the change in model output due to uncertainties in parameter values, is important.

There are certain characteristics that models have to varying degrees and which have a bearing on the question of how good they are:

- accuracy (is the output of the model correct?);
- descriptive realism (i.e. based on correct assumptions);
- precision (are predictions in the form of definite numbers?);
- robustness (i.e. relatively immune to errors in the input data);
- generality (applicable to a wide variety of situations);
- fruitfulness (a model is considered fruitful if its conclusions are useful or if it inspires development of other good models).

Step 7: Interpretation/application
The validated model is then ready to be used for one or several purposes as described earlier, e.g. to enhance our understanding, make predictions, and give information about how to control the process.

Let us conclude this chapter with a classical modeling problem attributable to Galileo Galilei (1564–1642).

Example 1.3 Galileo´s gravitation models
One of the oldest scientific investigations was the attempt to understand gravity. This problem provides a nice illustration of the steps in modeling.

"Understanding" gravity is too vague and ambitious a goal. A more specific question about gravity is:

Why do objects fall to the earth?

Aristotle´s answer was that objects fall to the earth because that is their natural place, but this never led to any useful science or mathematics. Around the time of Galileo (early seventeenth century), people began asking how gravity worked instead of why it worked. For example, Galileo wanted to describe the way objects gain velocity as they fall. One particular question Galileo asked was:

What relation describes how a body gains velocity as it falls?

The next step is to identify relevant factors. Galileo decided to take into account only distance, time, and velocity. However, he might have also considered the weight, shape, and density of the object as well as air conditions.

The first assumption Galileo made was:

Assumption 1 If a body falls from rest, its velocity at any point is proportional to the distance already fallen.

The mathematical description of Assumption 1 is:

$$\frac{dx}{dt} = ax. \tag{1.1}$$

This equation has the solution

$$x = ke^{at}. \tag{1.2}$$

The constant k is evaluated by

$$x(0) = 0, \tag{1.3}$$

giving $k = 0$, and thus

$$x = 0 \text{ for all } t.$$

The implication is that the object will never move, no matter how long we wait!

Since this conclusion is clearly absurd, and there are no mistakes in the mathematical manipulation, the model has to be reformulated. Galileo eventually came to this conclusion, and replaced Assumption 1 with:

Assumption 2 If a body falls from rest, its velocity at any point is proportional to the time it has been falling.

The mathematical description of this assumption is:

$$\frac{dx}{dt} = bt, \tag{1.4}$$

and the solution, with $x(0) = 0$, is

$$x = bt^2. \tag{1.5}$$

This law of falling bodies agrees well with observations in many circumstances, and the parameter b can be estimated from matching experimental data. Incidentally, the model constant b equals the gravitational constant, g.

1.3   Questions

(1) Give some reasons for doing mathematical modeling in chemical engineering.

(2) Explain why the model development often becomes an iterative procedure.

2 Classification

In this chapter mathematical models are classified by

- grouping into opposite pairs;
- mathematical complexity;
- degree of resolution.

The intention is to give the reader an understanding of differences between models as reflected by the modeling goal. Which question is the model intended to answer?

2.1 Grouping of models into opposite pairs

In this section, we will examine various types of mathematical models. There are many possible ways of classification. One possibility is to group the models into opposite pairs:

- linear versus non-linear;
- steady state versus non-steady state;
- lumped parameter versus distributed parameter;
- continuous versus discrete variables;
- deterministic versus stochastic;
- interpolation versus extrapolation;
- mechanistic versus empirical;
- coupled versus not coupled.

Linear versus non-linear
Linear models exhibit the important property of superposition; non-linear ones do not. Equations (and thus models) are linear if the dependent variables or their derivatives appear only to the first power; otherwise they are non-linear. In practice, the ability to use a linear model for a process is of great significance. General analytical methods for equation solving are all based on linearity. Only special classes of non-linear models can be attacked with mathematical methods. For the general case, where a numerical method is required, the amount of computation is also much less for linear models, and in addition error estimates and convergence criteria are usually derived under linear assumptions.

Steady state versus transient

Other synonyms for steady state are time invariant, static, or stationary. These terms refer to a process in which the point values of the dependent variables remain constant over time, as at steady state and at equilibrium. Non-steady-state processes are also called unsteady state, transient, or dynamic, and represent a situation in which the process dependent variables change with respect to time. A typical example of a non-steady-state process is the start-up of a distillation column which would eventually reach a pseudosteady-state set of operating conditions. Inherently transient processes include fixed-bed adsorption, batch distillation, reactors, drying, and filtration/ sedimentation.

Lumped parameter versus distributed parameter

A lumped-parameter representation means that spatial variations are ignored, and the various properties and the state of a system can be considered homogeneous throughout the entire volume. A distributed-parameter representation, in contrast, takes into account detailed variations in behavior from point to point throughout the system. All real systems are, of course, distributed in that some variations occur throughout them. As the variations are often relatively small, they may be ignored, and the system may then be "lumped."

The answer to the question whether or not lumping is valid for a process model is far from simple. A good rule of thumb is that if the response of the process is "instantaneous" throughout the process, then the process can be lumped. If the response shows instantaneous differences throughout the process (or vessel), then it should not be lumped. Note that the purpose of the model affects its validity. Had the purpose been, for example, to study mixing in a stirred tank reactor, a lumped model would be completely unsuitable because it has assumed from the first that the mixing is perfect and the concentration a single variable.

Because the mathematical procedures for solving lumped-parameter models are simpler than those for solving distributed-parameter models, we often approximate the latter using an equivalent lumped-parameter system. Whilst lumping is often possible, we must be careful to avoid masking the salient features of a distributed element and subsequently building an inadequate model by lumping.

As an example of the use of lumped versus distributed mathematical models, consider the equilibrium stage concept of distillation, extraction, and similar processes. As shown in Figure 2.1, we usually assume that the entire stage acts as a whole, and we do not consider variations in temperature, composition, or pressure in various parts of the stage. All of these variables are "lumped" together into some overall average. The errors introduced are compensated for by the stage efficiency factor.

Continuous versus discrete variables

Continuous means that the variables can assume any values within an interval; discrete means that a variable can take on only distinct values within an interval. For example, concentrations in a countercurrent packed bed are usually modeled in terms of continuous variables, whereas plate absorbers are modeled in terms of staged multicompartment

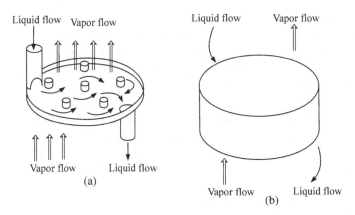

Figure 2.1. Lumped-parameter and distributed-parameter visualization of a distillation tray. (a) Actual plate with complex flow patterns and resulting variations in properties from point to point. (b) Idealized equilibrium stage ignoring all internal variations.

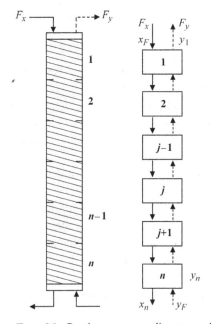

Figure 2.2. Continuous versus discrete modeling of a packed column absorber.

models in which a concentration is uniform at each stage but differs from stage to stage in discrete jumps. Continuous models are described by differential equations and discrete models by difference equations. Figure 2.2 illustrates the two configurations.

The left-hand figure shows a packed column modeled as a continuous system, whereas the right-hand figure represents the column as a sequence of discrete (staged) units. The concentrations in the left-hand column would be continuous variables; those in the right-hand column would involve discontinuous jumps. The tick marks in the left-hand column represent hypothetical stages for analysis. It is, of course, possible to model the packed

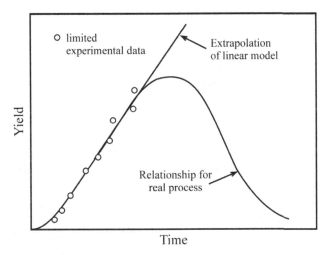

Figure 2.3. Danger of extrapolation. Yield of a chemical reactor versus time.

column in terms of imaginary segregated stages and to treat the plate column in terms of partial differential equations in which the concentrations are continuous variables.

Deterministic versus stochastic

Deterministic models or elements of models are those in which each variable and parameter can be assigned a definite fixed number, or a series of numbers, for any given set of conditions, i.e. the model has no components that are inherently uncertain. In contrast, the principle of uncertainty is introduced in stochastic or probabilistic models. The variables or parameters used to describe the input–output relationships and the structure of the elements (and the constraints) are not precisely known. A stochastic model involves parameters characterized by probability distributions. Due to this the stochastic model will produce different results in each realization.

Stochastic models play an important role in understanding chaotic phenomena such as Brownian motion and turbulence. They are also used to describe highly heterogeneous systems, e.g. transport in fractured media. Stochastic models are used in control theory to account for the irregular nature of disturbances. In the present context, we will focus upon deterministic models.

Interpolation versus extrapolation

A model based on interpolation implies that the model is fitted to experimentally determined values at different points and that the model is used to interpolate between these points. A model used for extrapolation, in comparison, goes beyond the range of experimental data.

Typically, thermodynamic models are used for interpolation as well as correlations in complicated transport phenomena applications. Extrapolation requires, in general, a detailed mechanistic understanding of the system. The procedure requires great care to avoid misleading conclusions. Figure 2.3 illustrates an exaggerated case of extrapolation by means of a linear model into a region beyond the range of experimental data for a chemical reaction that reaches a maximum yield in time.

In the safety analysis of nuclear waste repositories models are used to predict the fate of leaking radionuclides into the surrounding rock formation over geological time scales. Naturally, it is of the utmost importance that these models are physically/chemically sound and based on well-understood mechanistic principles.

Mechanistic versus empirical

Mechanistic means that models are based on the underlying physics and chemistry governing the behavior of a process; empirical means that models are based on correlated experimental data. Empirical modeling depends on the availability of process data, whereas mechanistic modeling does not; however, a fundamental understanding of the physics and chemistry of the process is required. Mechanistic models are preferably used in process design, whereas empirical models can be used when only trends are needed, such as in process control. Semi-empirical models cover the range in between. This discussion closely resembles the one regarding extrapolation/interpolation.

Coupled versus not coupled

When a model consists of two or more interacting relations, we have a coupled model. The coupling may be weak or strong. If the interaction only works in one direction, we speak of weak coupling (one-way coupling); if it operates in both directions we speak of strong (two-way) coupling. Forced convection is an example of one-way coupling, and free convection is an example of two-way coupling. In forced convection, the flow field is independent of the transport of energy and can be solved first and then introduced into the energy equation. In free convection, flow and energy transport are intimately coupled since the flow is generated by density differences originating from temperature differences. A model of a pneumatic conveying dryer involves a high degree of coupling (see Example 1.1).

2.2   Classification based on mathematical complexity

Another classification scheme to be considered is shown in Figure 2.4. It can be seen that the complexity of solving a mathematical problem roughly increases as we go down Figure 2.4. In other words, algebraic equations are usually easier to solve than ordinary differential equations, which in turn are usually easier to solve than partial differential equations. This is not always true, of course, since a linear partial differential equation may be easier to solve than a non-linear ordinary differential equation. The accuracy of the representation of the actual physical system attained using the mathematical model also roughly increases as we go down the figure, because the more independent variables and parameters that are taken into account, the more accurate the mathematical model will be.

The theory of ordinary differential equations is reasonably well advanced with regard to analytical solutions, but the same is not true for the theory of partial differential

Table 2.1. Classification of mathematical problems and their ease of solution using analytical methods

Equation	Linear equations			Non-linear equations		
	One equation	Several equations	Many equations	One equation	Several equations	Many equations
Algebraic	trivial	easy	essentially impossible	very difficult	very difficult	impossible
Ordinary differential	easy	difficult	essentially impossible	very difficult	impossible	impossible
Partial differential	difficult	essentially impossible	impossible	essentially impossible	impossible	impossible

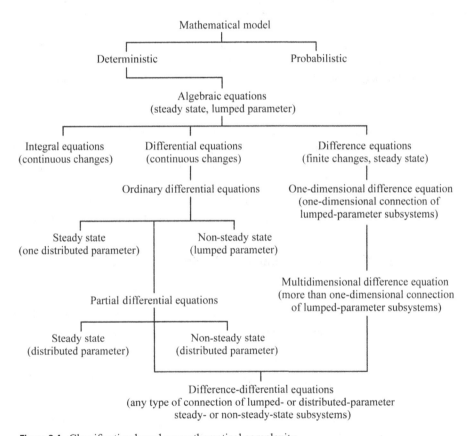

Figure 2.4. Classification based on mathematical complexity.

equations. Thus we can rather seldom find the analytical solution to a partial differential equation, and, in fact, when we do, it very often involves such things as infinite series, which are sometimes difficult to handle computationally. Table 2.1 shows the various classes of mathematical equations and the limited class amenable to analytical solution.

It should be noted that, in a model with more than one equation, the difficulty in obtaining a solution is dependent on the degree of coupling.

Table 2.2. Classification of models according to scale

Level of physicochemical description	Topical designations	Parameters
Molecular	treats discrete entities; quantum mechanics, statistical mechanics, kinetic theory	distribution functions; collision integrals
Microscopic	laminar transport phenomena, statistical theories of turbulence	phenomenological coefficients; viscosity, thermal conductivity, diffusivity
Mesoscopic	laminar and turbulent transport phenomena; transport in porous media	"effective" transport coefficients
Macroscopic	process engineering, unit operations	interphase transport coefficients

A model formulated in terms of differential equations can often be rephrased in terms of integral equations (and vice versa) so that many additional models are essentially included in this classification scheme. Difference equations account for finite changes from one stage to another and have significance parallel to that given above for (continuous) differential equations.

The classification scheme given in Table 2.1 for *analytical* methods has its counterpart for *numerical* methods. In such a case, the borderline to difficult/impossible problems is shifted to the right. In most cases, the models need to be solved numerically. Some reasons for this might be non-linearities, varying material properties, and varying boundary conditions. Luckily the computational power available in modern computers seldom conflicts with the requirement of solving the model equations numerically.

2.3 Classification according to scale (degree of physical detail)

Physicochemical models based on the degree of internal detail of the system encompassed by the model are classified in Table 2.2. The degree of detail about a process decreases as we proceed down the table.

Molecular description

The most fundamental description of processes, in the present context, would be based on molecular considerations. A molecular description is distinguished by the fact that it treats an arbitrary system as if it were composed of individual entities, each of which obeys certain rules. Consequently, the properties and state variables of the system are obtained by summing over all of the entities. Quantum mechanics, equilibrium and non-equilibrium statistical mechanics, and classical mechanics are typical methods of analysis, by which the properties and responses of the system can be calculated.

Microscopic description

A microscopic description assumes that a process acts as a continuum and that the mass, momentum, and energy balances can be written in the form of phenomenological equations. This is the "usual" level of transport phenomena where detailed molecular

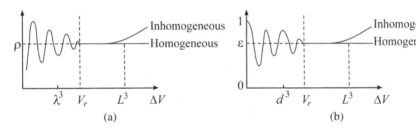

Figure 2.5. Concept of representative elementary volume for a fluid (a) and a porous medium (b), respectively.

interactions are ignored and differential balance equations are formulated for momentum, energy, and mass.

The continuum concept is illustrated in Figure 2.5(a) with the density of a fluid. The density, ρ, at a particular point in the fluid is defined as

$$\rho = \lim_{\Delta V \to V_r} \frac{\Delta m}{\Delta V}, \tag{2.1}$$

where Δm is the mass contained in a volume ΔV, and V_r is the smallest volume (the representative elementary volume) surrounding the point for which statistical averages are meaningful (in the figure, λ is the molecular mean free path and L is the macroscopic length scale). For air at room temperature and atmospheric pressure, the mean free path, λ, is approximately 80 nm. The concept of the density at a mathematical point is seen to be fictitious; however, taking $\rho = \lim_{\Delta V \to V_r}(\Delta m/\Delta V)$ is extremely useful, as it allows us to describe the fluid flow in terms of continuous functions. Note that, in general, the density may vary from point to point in a fluid and may also vary with respect to time.

Mesoscopic description
The next level of description, mesoscopic, involves averaging at larger scales and thus incorporates less detailed information about the internal features of the system of interest. This level is of particular interest for processes involving turbulent flow or flow in geometrically complex systems on a fine scale, such as porous media. The values of the dependent variables are averaged in time (turbulence) or space (porous media). Processes at this level are described by "effective" transport coefficients such as eddy viscosity (turbulence) or permeability (porous media).

The continuum concept at the porous media level is illustrated in Figure 2.5(b) for porosity:

$$\varepsilon = \lim_{\Delta V \to V_r} \frac{\Delta V_v}{\Delta V}, \tag{2.2}$$

where ΔV_v is the void volume in ΔV, and d is the pore length scale.

Time averaging in turbulence is illustrated in Figure 2.6. The instantaneous velocity v_z oscillates irregularly. We define the time-smoothed velocity $\bar{v}_z$ by taking a time average of v_z over a time interval t_0, which is large with respect to the time of turbulent oscillation

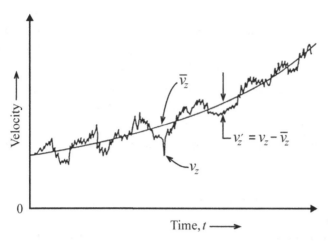

Figure 2.6. Time averaging.

but small with respect to the overall time changes:

$$\overline{v}_z = \frac{1}{t_0} \int_t^{t+t_0} v_z \, dt. \tag{2.3}$$

Macroscopic description

The final level, macroscopic, ignores all the details within a system and merely creates a balance equation for the entire system. The dependent variables, such as concentration and temperature, are not functions of position, but represent overall averages throughout the volume of the system. The model is effective as long as detailed information internal to the system is not required in model building. Macroscopic and lumped mean the same thing.

In science, events distinguished by large differences in scale often have very little influence on one another. The phenomena can, in such a case, be treated independently. Surface waves in a liquid, for instance, can be described in a manner that ignores the molecular structure of the liquid. Almost all practical theories in physics and engineering depend on isolating a limited range of length scales. This is why the kinetic theory of gases ignores effects with length scales smaller than the size of a molecule and much larger than the mean free path of a molecule. There are, however, some phenomena where events at many length scales make contributions of equal importance. One example is the behavior of a liquid near the critical point. Near that point water develops fluctuations in density at all possible scales: drops and bubbles of all sizes occur from single molecules up to the volume of the specimen.

2.4   Questions

(1) What is the difference between a lumped- and a distributed-parameter model?

(2) Explain the difference between deterministic and stochastic models.

(3) Why is a linear mathematical model tractable for analytical solution?

(4) Describe the continuum concept.

(5) At what scale is Darcy's law formulated? Are there alternatives that describe flow in porous materials?

3 Model formulation

Formulating mathematical models by applying balance and conservation principles and constitutive relations for fluxes is the topic of this chapter. The aim is to give the reader tools and skills for:

- constructing models using balances on differential or macroscopic control volumes for momentum, heat, mass, and numbers (population balances);
- constructing models by simplifying general model equations.

3.1   Balances and conservation principles

Before formulating a model it is crucial to define the system boundary. The purpose of the boundary is to define the system in relation to its surroundings. In Figure 3.1, a stirred tank is isolated from its surroundings by the dashed circle. All significant phenomena enclosed within this boundary need to be included in a successful model. The system boundary may be chosen in different ways, but for most systems the boundary to use is natural. Models derived from physicochemical principles are usually based on the general balance concept:

$$\begin{bmatrix} accumulation \\ within\ system \end{bmatrix} = \begin{bmatrix} net\ transport\ into \\ system\ through \\ boundaries \end{bmatrix} + \begin{bmatrix} net\ generation \\ within\ system \end{bmatrix}.$$

This relation is very general. The objective of model building is to transform the verbal concept into mathematical statements that are specific to the quantity of interest. We may balance mass, energy, and momentum as well as, for example, entropy and countable entities such as size and age distributions (population balances). Some of these entities are conserved, for example total mass, whilst some are not, for example the mass of a species in a mixture (due to chemical reactions).

By using the balance principle, we can derive model equations by balancing the quantities within the defined system boundary. A few examples of important balance equations are given in the following.

Overall (total) mass balance
The overall total mass balance describes the total mass in a system. Obviously there can be only one total mass balance equation, and the net generation term is zero, which means that mass is a conserved quantity.

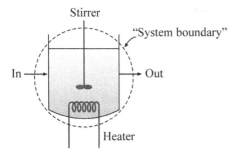

Figure 3.1. System boundary for a stirred tank.

Species (component) material balance
Most chemical engineering systems contain more than one component. In contrast to the total mass, the components or species are not always conserved; they may be generated or consumed due to chemical reactions. In this case, the net generation rate within the system must be quantified by reaction rate equations.

An N-component system will have N balance equations. As the species balance equations and the total balance are related, the $N-1$ species balances and the overall mass balance are sufficient to describe completely the mass flow in the system.

Energy balance
An energy balance equation for the total energy, TE, in the system, taking into consideration temperature-dependent thermal energy U, the potential energy PKE, pressure energy PE, and the kinetic energy KE, is given by

$$\begin{bmatrix} accumulation \\ of\ TE \\ within\ system \end{bmatrix} = \begin{bmatrix} net\ flow\ of\ U, PKE, \\ PE,\ KE\ into\ system \\ through\ boundaries \end{bmatrix} + \begin{bmatrix} net\ generation \\ within \\ system \end{bmatrix}.$$

In general, transformation between all these energies must be considered. In some cases, the transformation of mechanical energy, i.e. between PKE, PE, and KE, is of primary interest, and isothermal conditions may be assumed. In other cases, the potential and kinetic energies can be neglected, because the difference in elevation and low velocities contribute to small changes in total energy. Consequently, it is often possible to neglect the transformation of potential and kinetic energies to thermal energy, and the energy balance simplifies to a balance equation for the thermal energy.

As an example, consider a pipe flow where the fluid is being heated externally to 20 °C and is simultaneously decelerated due to an increase in the pipe diameter. In this case, the kinetic energy will typically contribute little to temperature change in comparison to the heating. Let us assume that the velocity at the inlet equals 1 m s^{-1}. The limiting case with a maximum temperature increase occurs with infinite pipe expansion, where the velocity reduces to 0 m s^{-1}. The balance between enthalpy and kinetic energy is given by $\rho c_p \Delta T = \rho \Delta v^2/2$. For water, a fluid with a high heat capacity, approximately 4.2 J g^{-1} K^{-1}, the temperature increase will be $\Delta T \approx 0.1$ °C. Consequently, the kinetic

Table 3.1. Constitutive relations in transport phenomena models

Mechanism	Dimension	Flux	Name
Momentum	N m^{-2}	$\tau_{yx} = -\mu \dfrac{\partial v_x}{\partial y}$	Newton
Heat	J m^{-2} s^{-1}	$q_x = -k \dfrac{\partial T}{\partial x}$	Fourier
Mass	mol m^{-2} s^{-1}	$J_{A,x} = -C D_{AB} \dfrac{\partial y_A}{\partial x}$	Fick
Flow (porous)	m^{-3} m^{-2} s^{-1}	$q_x = -\dfrac{k}{\mu} \dfrac{\partial P}{\partial x}$	Darcy

energy leads to an almost negligible increase in temperature. In this heating application, the balance equation can be reduced safely to a thermal energy balance.

Because the thermal energy not only depends on temperature, but also on the composition, the energy balance equation is most often accompanied by species (material) balance equations. Several phenomena may cause heat generation within a system, e.g. reaction heat (exothermic and endothermic reactions), heating and cooling due to convection, conduction, radiation, condensation, evaporation, work done by the system on the surroundings, and shaft work due to an impeller.

Momentum balance
Momentum and force are vector quantities, and the number of equations equals the number of spatial dimensions in the model, e.g. a 2D model must contain momentum balances for x- and y-momentum. Three kinds of forces are typically accounted for in chemical engineering: pressure force, shear force, and gravitational force. A force is associated with momentum production (Newton's second law) and thus enters via the last term in the general balance.

3.2   Transport phenomena models

The most common types of models in chemical engineering are those related to the transport of mass, heat, and momentum. In addition to the balance equation, a constitutive equation that relates the flux of interest to the dependent variable (e.g. mass flux to concentration) is needed. These relations (in simple 1D form) for the microscopic level and for flow at the porous media level are given in Table 3.1. It should be noted that all these relations have the general form

$$\text{flux} = \text{transport coefficient} \times \text{gradient}.$$

A simple example of setting up a model at the microscopic level (1D transient heat conduction with a heat source) is given in Example 3.1 (Figure 3.2).

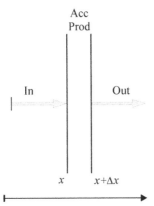

Figure 3.2. Simple shell balance.

Example 3.1 1D transient heat conduction with a source term

The variables and parameters used in this example are as follows: ρ = density, c_p = heat capacity, q = heat flux, k = thermal conductivity, A = cross-sectional area, S = source strength (J m^{-3} s^{-1}).

accumulation:	$\rho c_p \Delta T (A \Delta x)$;
in (conduction):	$(qA)|_x \Delta t$;
out (conduction):	$(qA)|_{x+\Delta x} \Delta t$;
production:	$S A \Delta x \Delta t$.

Insertion of these terms into the balance equation yields

$$\rho c_p \frac{\Delta T}{\Delta t} = \frac{q|_x - q|_{x+\Delta x}}{\Delta x} + S.$$

By taking the limit as $\Delta x \to 0$ and $\Delta t \to 0$ we have

$$\rho c_p \frac{\partial T}{\partial t} = -\frac{\partial q}{\partial x} + S.$$

Finally, introducing Fourier's law, $q_x = -k(\partial T/\partial x)$, yields

$$\rho c_p \frac{\partial T}{\partial t} = k \frac{\partial^2 T}{\partial x^2} + S,$$

that is to be solved with proper boundary and initial conditions.

General microscopic transport equations can be found in standard text books on transport phenomena (see the Bibliography). For example, the transient 3D binary mass transport equation in rectangular coordinates is obtained by using a differential control volume in three dimensions, as depicted in Figure 3.3.

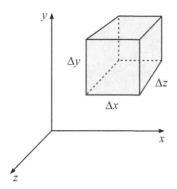

Figure 3.3. Differential control volume in three dimensions.

Example 3.2 Derivation of the transient 3D binary mass transport equation in rectangular coordinates

The variables and parameters used in this example are as follows: C_A = concentration, J_A = diffusion flux, v = velocity, R_A = source term (mol m^{-3} s^{-1}).

accumulation:	$\Delta C_A \Delta x \Delta y \Delta z$;
in (diffusion):	x-direction $J_{A,x}\|_x \Delta y \Delta z \Delta t$,
	y-direction $J_{A,y}\|_y \Delta x \Delta z \Delta t$,
	z-direction $J_{A,z}\|_z \Delta x \Delta y \Delta t$;
in (flow):	x-direction $(v_x C_A)\|_x \Delta y \Delta z \Delta t$,
	y-direction $(v_y C_A)\|_y \Delta x \Delta z \Delta t$,
	z-direction $(v_z C_A)\|_z \Delta x \Delta y \Delta t$;
out (diffusion):	x-direction $J_{A,x}\|_{x+\Delta x} \Delta y \Delta z \Delta t$,
	y-direction $J_{A,y}\|_{y+\Delta y} \Delta x \Delta z \Delta t$,
	z-direction $J_{A,z}\|_{z+\Delta z} \Delta x \Delta y \Delta t$;
out (flow):	x-direction $(v_x C_A)\|_{x+\Delta x} \Delta y \Delta z \Delta t$,
	y-direction $(v_y C_A)\|_{y+\Delta y} \Delta x \Delta z \Delta t$,
	z-direction $(v_z C_A)\|_{z+\Delta z} \Delta x \Delta y \Delta t$;
production:	$R_A \Delta x \Delta y \Delta z \Delta t$.

Using the balance equation and taking the limit as $\Delta x, \Delta y, \Delta z, \Delta t \rightarrow 0$ yields

$$\frac{\partial C_A}{\partial t} = -\frac{\partial(v_x C_A)}{\partial x} - \frac{\partial(v_y C_A)}{\partial y} - \frac{\partial(v_z C_A)}{\partial z} - \frac{\partial J_{A,x}}{\partial x} - \frac{\partial J_{A,y}}{\partial y} - \frac{\partial J_{A,z}}{\partial z} + R_A. \quad (3.1)$$

Finally, using Fick's law, $J_{A,x} = -D_{AB}(\partial C_A/\partial x)$ etc., we obtain

$$\frac{\partial C_A}{\partial t} = -\frac{\partial(v_x C_A)}{\partial x} - \frac{\partial(v_y C_A)}{\partial y} - \frac{\partial(v_z C_A)}{\partial z} + D_{AB}\frac{\partial^2 C_A}{\partial x^2} + D_{AB}\frac{\partial^2 C_A}{\partial y^2}$$

$$+ D_{AB}\frac{\partial^2 C_A}{\partial z^2} + R_A, \quad (3.2)$$

or, in vector notation,

$$\frac{\partial C_A}{\partial t} = -\nabla \cdot (vC_A) + D_{AB}\nabla^2 C_A + R_A. \tag{3.3}$$

For convenience, the general equations for mass, momentum, and heat transport, in rectangular, cylindrical, and spherical coordinates, are provided in Appendix A.

Given these general equations, one approach to mathematical modeling is to apply appropriately and simplify these equations for the particular problem. This involves a clear statement of assumptions and an estimation of the relative importance of various terms (e.g. using dimensional analysis; see Chapter 4). In Example 3.3, the energy balance equation has been simplified from its generic form to an equation describing the heat conduction in one dimension (Example 3.1).

Example 3.3 Reduction of the general energy balance in Example 3.1

The energy balance equation for rectangular coordinates is given by (see Appendix A)

$$\rho c_p \left(\frac{\partial T}{\partial t} + v_x \frac{\partial T}{\partial x} + v_y \frac{\partial T}{\partial y} + v_z \frac{\partial T}{\partial z} \right) = k \left(\frac{\partial^2 T}{\partial x^2} + \frac{\partial^2 T}{\partial y^2} + \frac{\partial^2 T}{\partial z^2} \right) + S.$$

In Example 3.1, there is no convective flow, consequently there is no heat convection, and the following terms are canceled:

$$v_x \frac{\partial T}{\partial x}, v_y \frac{\partial T}{\partial y}, v_z \frac{\partial T}{\partial z} \equiv 0.$$

Furthermore, it is a 1D problem, which means that conduction in the y- and z-directions can also be omitted, i.e.

$$k \frac{\partial^2 T}{\partial y^2}, k \frac{\partial^2 T}{\partial z^2} \equiv 0.$$

The only terms that remain in the general balance equation are accumulation, conduction in the x-direction, and the source term. Thus, the balance equation simplifies to

$$\rho c_p \frac{\partial T}{\partial t} = k \frac{\partial^2 T}{\partial x^2} + S,$$

which is the same model equation as derived in Example 3.1.

In Example 3.1, the flow was set to zero by definition. In many situations the problem is more complex and involves estimating the various terms. For instance, in Example 3.3 there may be both flow and conduction of heat. To estimate if one mechanism dominates, a dimensionless number comparing the terms can be used.

The total flux of a quantity is the sum of its molecular flux and convective flux. The ratio of the convective flux to the molecular flux (the Péclet number) can be used to determine the relative importance of each flux. The dimensionless Péclet numbers are defined by

$$Pe_{heat} = uL/\alpha \qquad \text{(heat transfer)},$$

$$Pe_{mass} = uL/D_{AB} \qquad \text{(mass transfer)}.$$

Here, u and L are the characteristic velocity and length scales, and α and D_{AB} are the diffusivities for heat and mass transport, respectively. They obviously both have the same unit ($\text{m}^2 \text{ s}^{-1}$). Heat diffusivity is defined as $\alpha = \lambda/(\rho c_p)$ where λ is the heat conductivity.

Depending on the magnitude of the Péclet number, we have

$$Pe \ll 1 \quad \text{total flux} \approx \text{molecular flux},$$

$$Pe \approx 1 \quad \text{total flux} = \text{molecular flux} + \text{convective flux},$$

$$Pe \gg 1 \quad \text{total flux} \approx \text{convective flux}.$$

3.3   Boundary conditions

Equally as important as formulating the differential equation(s) when developing a mathematical model is the selection of an appropriate set of boundary conditions and/or initial conditions. In order to calculate the values of arbitrary constants that evolve in the solution of a differential equation, we generally need a set of n boundary conditions for each nth-order derivative with respect to a space variable or with respect to time. For example, the differential equation

$$\rho c_p \frac{\partial T}{\partial t} = k \frac{\partial^2 T}{\partial x^2} + S \tag{3.4}$$

requires the value of T to be specified at two locations of x and one value of t.

Appropriate boundary conditions arise from the actual process or the problem statement. They essentially are given, or, more often, must be deduced from, physical principles associated with the problem. These physical principles are usually mathematical statements that show that the dependent variable at the boundary is at equilibrium, or, if some transport is taking place, that the flux is conserved at the boundary. Another type of boundary condition uses interfacial transport coefficients (e.g. heat transfer or mass transfer coefficients) that express the flux as the product of the interphase transport coefficient and some kind of driving force.

The common boundary conditions for use with momentum, energy, and mass transport are listed in Tables 3.2–3.4. Note the similarities among the three modes of transport. These boundary conditions apply to all strata of description shown in Table 2.2, except for the molecular one.

Recall the mathematical classification of boundary conditions summarized in Table 3.5. For example, in energy transport, the first type corresponds to the specified temperature at the boundary; the second type corresponds to the specified heat flux at the boundary; and the third type corresponds to the interfacial heat transport governed by a heat transfer coefficient.

Table 3.2. Common boundary conditions for use with the transport of mass

Description	Math
Concentration at a boundary is specified	$C = C_0$
Mass flux across a boundary is continuous	$(n_i)_{x=0-} = (n_i)_{x=0+}$
Concentrations on both sides of a boundary are related functionally	$(C_i)_{x=0-} = f(C_i)_{x=0+}$
Convective mass (mole) flux at a boundary is specified	$(N_A)_{x=0} = k_c(C_{bulk} - C_{surface})$
Rate of reaction at a boundary is specified	$(N_A)_{x=0} = R_A$

Table 3.3. Common boundary conditions for use with the transport of momentum

Description	Math
Velocity at a boundary is specified	$v_{rel} = 0$ (no-slip condition at solid–fluid interface)
Momentum flux across a boundary is continuous	e.g. τ is continuous at a liquid–liquid interface
Velocity at a boundary is continuous	$(v)_{x=0-} = (v)_{x=0+}$
Momentum flux is specified	e.g. τ in liquid is approximately zero at gas–liquid interfaces (at low relative velocities)

Table 3.4. Common boundary conditions for use with the transport of energy

Description	Math
Temperature at a boundary is specified	$T = T_0$
Heat flux across a boundary is continuous	$(q_i)_{x=0-} = (q_i)_{x=0+}$
Temperature at a boundary is continuous	$(T)_{x=0-} = (T)_{x=0+}$
Convective heat flux at a boundary is specified	$(q)_{x=0} = h(T_{bulk} - T_{surface})$
Heat flux at a boundary is specified	$q = q_0$

Table 3.5. Classification of boundary conditions

Type	Description	Math
Dirichlet (first type)	specifies the value a solution must take at its boundary	$y(0) = \gamma_1$
Neumann (second type)	specifies the value the derivative of the solution must take at its boundary	$\left.\dfrac{\partial y}{\partial x}\right|_{x=0} = \gamma_1$
Robin (third type)	specifies a linear combination of the value of the function and the value of its derivatives at the boundary	$a_1 y + b_1 \dfrac{\partial y}{\partial x} = \gamma_1$

In setting up the domain to be modeled, all symmetries in the problem should be used to reduce the computational domain. (As long as this does not compromise the physics of the system, i.e. we do not want to restrict the solution.)

3.4   Population balance models

The basis of a population balance model is that the number of entities with some property in a system is a balanceable quantity. Properties include, among others, size, mass, and age.

There are many examples in the process industries for which discrete entities are created, destroyed, or changed in some way as a result of processing. A classical example concerns crystallization, where the size distribution of crystals and its evolution is of the highest relevance. In this application, the evolution of the crystal size distribution is predicted using population balance models and closures describing mechanisms such as nucleation, growth, and breakage. Granulation is another example; in this process, fine particles are bound together into larger granules. Applications include manufacturing of pharmaceuticals, detergents, and fertilizers. Consequently, population balance models serve as a tool to predict, control, and optimize the complex dynamics of these systems. Many biochemical processes also have characteristics that lend themselves to analysis via the population balance model. Other examples include flocculation for purifying drinking water, gas–liquid dispersions, and liquid–liquid extraction and reaction. The residence time distribution (RTD) theory is a special case of the general population balance.

Let us first, as an introduction, discuss a commonly used population balance model for flocculation in stirred reactor tanks. The aim of flocculation is to agglomerate fine particles in water, using chemical additives, to large aggregates that are easy to separate in sedimentation processes. The agglomerates are formed due to binary collisions of particles. Not every collision is successful, however, so the collision efficiency has to be accounted for. As the agglomerates grow, there is an increasing risk that they fragment into smaller aggregates or even "primary" particles. This may be the result of shear forces or collisions with the impeller, walls, or other particles.

Figure 3.4 shows the experimental results, using laser techniques, to follow the evolution of floc sizes over time. The primary data have been evaluated, using image analysis techniques, and the result is also shown in the figure as the evolution of the number concentration of flocs of different sizes ("population") over time. It can be seen that the number of small flocs decrease, and larger flocs form, over time.

The population balance, including these effects, can be written as follows:

$$\frac{dn_k}{dt} = \frac{1}{2} \sum_{i+j=k} \alpha(i, j)\beta(i, j)n_i n_j - \sum_{i=1}^{\infty} \alpha(i, k)\beta(i, k)n_i n_k - \chi(k)n_k + \sum_{i>k}^{\infty} \chi(i)n_i.$$

$$(3.5)$$

In this equation, the term on the left-hand side represents the rate of change of the number concentration of agglomerates of size k (valid for any size k). This is the result of the processes accounted for on the right-hand side of the equation. The first term accounts for the formation of agglomerates of size k due to collisions of two smaller

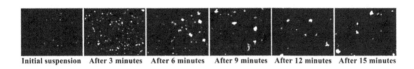

Initial suspension   After 3 minutes   After 6 minutes   After 9 minutes   After 12 minutes   After 15 minutes

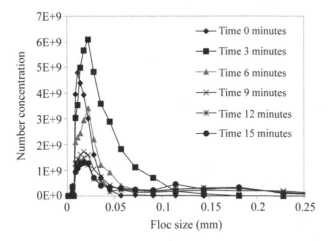

Figure 3.4. Change in floc size distribution during a flocculation process (Pelin, K., Licentiate thesis, Chalmers University of Technology, 1999).

particles with sizes i and j (the factor of ½ is to avoid counting the same collision twice). The second term gives a decrease in the number of k agglomerates, due to collisions with existing agglomerates of this size and other particles with arbitrary size i (note the negative sign). The third term, also negative, is a breakage function for particles of size k; and the last term is a breakage function for agglomerates larger than k giving fragments of size k. Here

$\alpha(i, j)$ is the collision efficiency of binary collision between i and j (a number between 0 and 1);

$\beta(i, j)$ is the collision frequency between i and j (strongly dependent upon local flow conditions);

$\chi(i)$ is the breakage function.

This population balance model is able to reproduce the results in Figure 3.4.

Flocculation is an example of a *discrete* growth of particles. In other processes, notably crystallization, the growth is *continuous*. As a second, introductory, example, we derive the population balance for batch crystallization (well mixed), only accounting for continuous growth given by the growth rate:

$$v = \frac{dm}{dt},$$

where m is the crystal mass. The number distribution is now given by $f(m, t)$, where

$f(m, t)\Delta m$ represents the number of crystals with mass within the range

$m, m + \Delta m$ at time t.

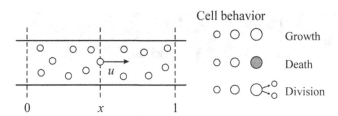

Figure 3.5. Cell behavior in a simple flow reactor.

The population balance for this case becomes

$$\Delta f \Delta m = [(fv)|_m - (fv)|_{m+\Delta m}]\Delta t.$$

The left-hand side of this relation represents the change in the number of crystals in the size range m, $m + \Delta m$ at time t. The right-hand side gives the number of smaller crystals that reach this size by continuous growth minus the number of crystals of the "right" size growing to a larger size.

Taking the limit as $\Delta m \Delta t \to 0$, we obtain

$$\frac{\partial f}{\partial t} = -\frac{\partial (fv)}{\partial m}, \tag{3.6}$$

which is to be solved with appropriate boundary conditions.

Before deriving the general population balances, we will give another example with one spatial dimension and one distributed property (mass of cell): continuous cell growth in a plug flow reactor.

Example 3.4 Population balance for cell behavior in a simple flow system

Let us consider a population of cells flowing through a plug flow reactor (Figure 3.5). The cells are characterized at time t by their position, x, and their mass, m. They are supposed to grow, to die, and to divide into two daughter cells (with mass conservation).

Figures 3.6(a) and (c) illustrate the "trajectories" of the cells in the physical space domain and in the mass–time domain, respectively. These two curves may be summarized in a single one, Figure 3.5(b), illustrating the mass–abscissa relationship. Figure 3.5(d) considers a small control surface Δx over Δm.

Let

$f(x, m, t)\Delta m \Delta x$ represent the number of cells within the range m, $m + \Delta m$ and x, $x + \Delta x$ at time t;

$G(x, m, t) = G^+ - G^-$ be the net generation of cells (where G^+ represents birth and G^- represents death);

$v (= dm/dt)$ be the growth rate of an individual cell in a uniform medium of constant composition.

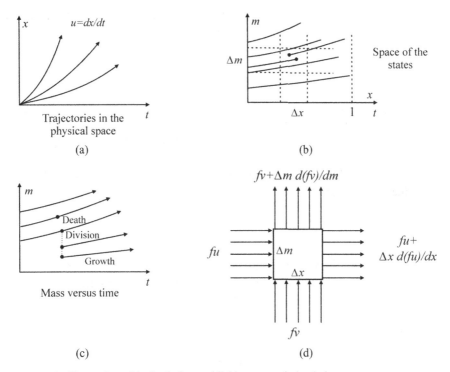

Figure 3.6. Illustration of the basis for establishing a population balance.

The mass balance for the cells in the (x, m, t) space is thus given by

$$\underbrace{\Delta f \, \Delta m \, \Delta x}_{\text{acc}} = \underbrace{[(fv)|_m - (fv)|_{m+\Delta m}] \Delta x \, \Delta t}_{\text{net "growth"}} + \underbrace{[(fu)|_x - (fu)|_{x+\Delta x}]}_{\text{net "inflow"}}$$

$$\times \Delta m \, \Delta t + \underbrace{G \Delta x \, \Delta m \, \Delta t.}_{\text{generation}}$$

Taking the limit as $\Delta m \, \Delta x \, \Delta t \to 0$, we obtain

$$\frac{\partial f}{\partial t} = -\frac{\partial (fv)}{\partial m} - \frac{\partial (fu)}{\partial x} + G. \tag{3.7}$$

This equation is readily generalized to three spatial dimensions by replacing the term $\partial(fu)/\partial x$ by $\nabla \cdot (fu)$.

There is a growing interest in what is, somewhat misleadingly, called "multidimensional" population balance models. One example of a "2D" population balance model is the description of a granulation process where not only the particle size distribution with time, but also the fractional binder content is predicted by the model. The binder (liquid) content of the granules governs the agglomeration process.

General 3D microscopic population balance equations, including several distributed properties, e.g. size and moisture content, can be derived using the same balance method.

Let us denote the distributed properties by p_i, where v_i is their respective rate of change ($v_i = dp_i/dt$). The local population balance equation then becomes

$$\frac{\partial f}{\partial t} = -\sum_{i=1}^{n} \frac{\partial(f v_i)}{\partial p_i} - \nabla \cdot (f u) + G. \tag{3.8}$$

In the case of granulation, there will now be two terms in the sum on the right-hand side of the equation: one associated with granule growth, and one associated with change in liquid content. Note that particle velocity may differ from the fluid velocity due to slip or external forces.

In cases where diffusion cannot be neglected, the additonal term $\nabla \cdot [D_p \nabla f]$ must be included in Equation 3.8:

$$\frac{\partial f}{\partial t} = -\sum_{i=1}^{n} \frac{\partial(f v_i)}{\partial p_i} - \nabla \cdot (f u) + \nabla \cdot [D_p \nabla f] + G. \tag{3.9}$$

Sometimes it is convenient to define a *macroscopic population balance* by averaging over the physical space and over the inlet and outlet streams. Let us define

$$\langle f \rangle = \frac{1}{V} \iiint_V f \, dV,$$

$$Q_{in} f_{in} = \iint_{S_{in}} u f \, dS,$$

$$Q_{out} f_{out} = \iint_{S_{out}} u f \, dS.$$

Integrating Equation (3.8) over space and dividing by V (note that V is not necessarily constant with time) yields

$$\frac{1}{V} \frac{\partial(V \langle f \rangle)}{\partial t} = -\sum_{i=1}^{n} \frac{\partial(\langle f v_i \rangle)}{\partial p_i} - \frac{1}{V}(Q_{out} f_{out} - Q_{in} f_{in}) + \langle G \rangle. \tag{3.10}$$

Equation (3.10) can, of course, be derived directly using a macroscopic control volume.

3.4.1   Application to RTDs

The concept of residence time distributions (RTDs) is central to classical chemical engineering. In this concept, the outlet response of a system to a known input disturbance (i.e. tracer concentration) is analyzed. Relations are derived for the outlet age distribution using material balances. In the following, we will demonstrate that identical relations are obtained via a completely different route using population balances.

Let us consider the age of fluid "particles" in an arbitrary flow system (no generation). If we denote age by α, the growth rate of age is given by

$$v = \frac{d\alpha}{dt} = 1. \tag{3.11}$$

Let f be the number of fluid particles of age α. Equation (3.10) for this case becomes

$$\frac{1}{V}\frac{\partial(V\langle f\rangle)}{\partial t} = -\frac{\partial\langle f\rangle}{\partial\alpha} - \frac{1}{V}(Q_{out}f_{out} - Q_{in}f_{in}). \tag{3.12}$$

Define the normalized distributions:

$$\langle f_n\rangle = \langle f\rangle/\langle C\rangle,$$
$$f_{in,n} = f_{in}/C_{in},$$
$$f_{out,n} = f_{out}/C_{out},$$

where $\langle C\rangle$, C_{in}, and C_{out} are the mean concentration, and the inlet and outlet concentrations, respectively.

Insertion of these definitions into Equation (3.12) and using the mass balance,

$$\frac{\partial(V\langle C\rangle)}{\partial t} = Q_{in}C_{in} - Q_{out}C_{out}, \tag{3.13}$$

yields

$$\frac{\partial\langle f_n\rangle}{\partial t} = -\frac{\partial\langle f_n\rangle}{\partial\alpha} - \langle f_n\rangle\left(\frac{1}{\tau_{in}^1} - \frac{1}{\tau_{out}^1}\right) - \left(\frac{f_{out,n}}{\tau_{out}^1} - \frac{f_{in,n}}{\tau_{in}^1}\right), \tag{3.14}$$

where τ^1 are characteristic hydrodynamic times defined on a molar (or mass) basis, and

$$\tau_{in}^1 = (V\langle C\rangle)/Q_{in}C_{in},$$
$$\tau_{out}^1 = (V\langle C\rangle)/Q_{out}C_{out}.$$

Using standard RTD nomenclature:

$\langle f_n\rangle$ is the normalized internal age distribution, $I(\alpha, t)$,
$f_{out,n}$ is the distribution of age in the outlet stream, $E(\alpha, t)$,
$f_{in,n}$ is the normalized distribution of age in the inlet stream, i.e. $\delta(\alpha)$ (Dirac's delta function).

Consequently, Equation (3.14) may be written as follows:

$$\frac{\partial I}{\partial t} = -\frac{\partial I}{\partial\alpha} - I\left(\frac{1}{\tau_{in}^1} - \frac{1}{\tau_{out}^1}\right) - \left(\frac{E}{\tau_{out}^1} - \frac{\delta(\alpha)}{\tau_{in}^1}\right). \tag{3.15}$$

Equation (3.15) is the generalized relationship between the residence time distribution, E, and the internal age distribution, I, in the transient state for an arbitrary flow system.

For an incompressible fluid (given a constant volume),

$$Q_{in} = Q_{out}.$$

Assuming further steady state:

$$C_{in} = C_{out}$$

and

$$\tau_{in}^1 = \tau_{out}^1 = \tau^1.$$

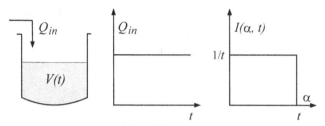

Figure 3.7. Internal age distribution in a semi-batch reactor.

Equation (3.15) simplifies to

$$\tau^1 \frac{dI}{d\alpha} + E = \delta(\alpha). \tag{3.16}$$

Example 3.5 Semi-batch reactor (incompressible)

Let us consider a single incompressible fluid that is poured into the reactor at a constant flow rate Q_{in} (see Figure 3.7). Since there is no outlet stream, $E = 1/\tau_{out}^1 = 0$. The inlet hydrodynamic time is given by

$$\tau_{in}^1 = V/Q_{in} = t Q_{in}/Q_{in} = t,$$

and Equation (3.12) reduces to

$$\frac{\partial I}{\partial t} = -\frac{\partial I}{\partial \alpha} - \frac{I}{t} + \frac{\delta(\alpha)}{t}.$$

The solution of this equation is

$$I(\alpha, t) = \frac{1}{t}(H(\alpha) - H(t - \alpha)),$$

where H is the Heaviside step function.

From this result, we can deduce the following:

- the ages are less than t (luckily!);
- the fraction of fluid particles of age between α and $\alpha + d\alpha$ is the volume $dV = Q_{in} \, d\alpha$ poured at $t - \alpha$ divided by the volume $V = Q_{in} t$ of fluid at time t.

Consequently,

$$I \, d\alpha = dV/V = d\alpha/t.$$

3.5   Questions

(1) What are the key steps in deriving a transport phenomena model?

(2) What does a generic balance equation describe?

(3) Explain what is meant by a boundary condition, and how they can be classified.

(4) What is the difference between a balance and a conservation principle?

(5) What is a population balance?

(6) What is the balanceable property in residence time distributions?

3.6   Practice problems

3.1 A spherical nuclear fuel element consists of a sphere of fissionable material with radius R_F, surrounded by a spherical shell of aluminum cladding with outer radius R_C. Fission fragments with very high kinetic energies are produced inside the fuel element. Collision between these fragments and the atoms of the fissionable material provide the major source of thermal energy in the reactor. The volume source of thermal energy, in the fissionable material, is assumed to be

$$S_n = S_{n0} \left(1 + b \left(\frac{r}{R_F} \right)^2 \right) \; (\mathrm{J\,m^{-3}\,s^{-1}}),$$

where b is a constant between 0 and 1, and $S_n = 0$ for $r > R_F$.

Derive a model for heat flux and temperature profile at steady conditions. Assume constant temperature T_0 at the outer surface.

3.2 It is possible to increase the energy transfer between a surface and an adjacent fluid by increasing the surface area in contact with the fluid. This may be accomplished by attaching metal fins to the surface (see Figure P3.1). Derive, using a shell balance, a stationary model for the heat transfer in a single fin. The wall temperature is T_0 and the surrounding fluid is at temperature T_∞. Radial temperature variations in the fin can be neglected, and the convective heat transfer coefficient, h, is given.

The cross-sectional area A and circumference P vary with distance from the wall according to $A = A(x)$ and $P = P(x)$.

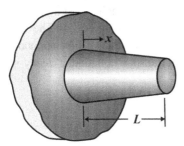

Figure P3.1

3.3 In the Liseberg Amusement Park in Gothenburg, there is a "fountain" where a water film flows along the outside of a vertical pipe. Derive a model for the velocity profile in the film:

(a) by using a shell balance approach;

(b) by simplifying the general momentum transport equation.

3.4 In making tunnels in the ground (porous material), water infiltration is a complicating factor. Assume a cylindrical tunnel (see Figure P3.2) with $P_1 > P_0$. Formulate a model that can be used to calculate the flow into the tunnel. Assume that Darcy's law is valid.

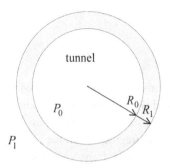

Figure P3.2

3.5 Around 1850, Fick conducted experiments that convinced him of the correctness of the diffusion equation. He dissolved salt crystals at the bottom of the experimental setups shown in Figure P3.3. The water in the top section was continually renewed to fix the concentration at zero. What concentration profiles did Fick obtain in the lower section at steady conditions?

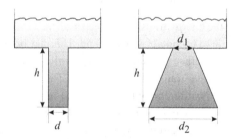

Figure P3.3

3.6 A fluid flows in the positive x-direction through a long flat duct of length L, width W, and thickness B, where $L \gg W \gg B$ (see Figure P3.4). The duct has porous walls at $y = 0$ and $y = B$, so that a constant cross flow can be maintained, with $v_y = v_0$. Derive, by simplifying the general transport equation, a steady-state model for the velocity distribution $v_x(y)$.

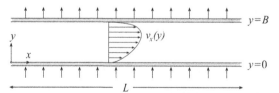

Figure P3.4

3.7 Consider two concentric porous spherical shells of radii κR and R (see Figure P3.5). The inner surface of the outer shell is kept at temperature T_1 and the outer surface of the inner shell is to be maintained at a lower temperature T_κ. Dry air (mass flow w_r kg s^{-1}) at temperature T_κ is blown radially from the inner shell into the intervening space and out through the outer shell.

Develop, using a shell balance, an expression for the temperature at steady conditions. In addition, give an expression for the rate of heat removal from the inner sphere. Assume steady laminar flow.

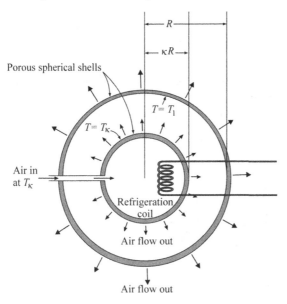

Figure P3.5

3.8 A liquid (B) is flowing at laminar conditions down a vertical wall (see Figure P3.6). For $z < 0$ the wall does not dissolve in the fluid, but for $0 < z < L$ the wall contains a species A that is slightly soluble in B. Develop a mathematical model for the dissolution process.

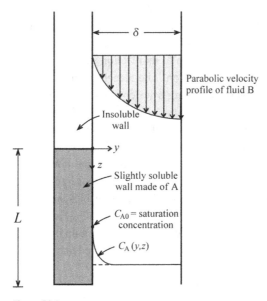

Figure P3.6

3.9 In a cookbook, a "gentle" way of preparing fish is suggested (a fish is ready when the temperature is around 60 °C). A pan is filled with water that is boiled. The pan is then taken from the stove and put upon a heat-insulating material. Thereafter the piece of fish is placed into the pan and the lid is put on. State a model for the process, assuming the following:

the piece of fish is "spherical" and initially at room temperature;
the water in the pan is well mixed.

3.10 Develop a model for the freezing of a spherical falling water drop (see Figure P3.7). The drop is surrounded by cool air at temperature T_∞ and its initial temperature is T_i (> freezing temperature T_0). Assume no volume change in the freezing process.

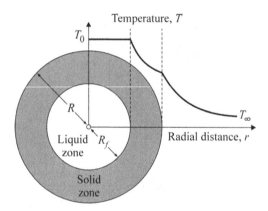

Figure P3.7

3.11 Chlorine dioxide (ClO_2) is a common chemical for bleaching aqueous suspensions of wood pulp. The ClO_2 reacts rapidly and irreversibly with lignin, which constitutes about 5% of the pulp; the remainder of the pulp (predominantly cellulose) is inert with respect to ClO_2. In addition to its reaction with lignin, ClO_2 also undergoes a slow spontaneous decomposition. The bleaching process may be studied under simplified conditions by assuming a water-filled mat of pulp fibers exposed to a dilute aqueous solution of ClO_2 (see Figure P3.8). The solution–mat interface is at $x = 0$ and the thickness of the mat is taken to be infinite. As ClO_2 diffuses in and reacts, the boundary between brown (unbleached) and white (bleached) pulp moves away from the interface, the location of that boundary being denoted by $x = \delta(t)$. Assuming the reaction between ClO_2 and lignin to be "infinitely" fast, a sharp moving boundary is formed separating bleached and unbleached pulp. The concentration of ClO_2 at $x = 0$ is constant, C_{A0}, and the unbleached lignin concentration is C_{B0}. Lignin is a component of the pulp fibers and may be regarded as immobile. The decomposition of ClO_2 is assumed to be as a first-order chemical reaction.

Develop a model of the process.

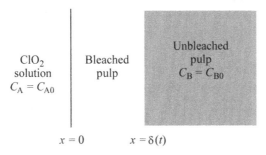

Figure P3.8

3.12 Erik released two newly born rabbits on the island of Tistlarna in the archipelago
of Gothenburg during the summer. Now it is autumn, and he wonders how many
rabbits there will be in five years.

Help Erik to solve this problem, assuming that rabbits are born pair-wise (one
female and one male) and that the number of pairs follows the Fibonacci sequence:
1, 1, 2, 3, 5, 8, ... (every number in the series is the sum of its two predecessors).
Assume further that the death frequency is given by

$$death\ frequency\ (age) = a \cdot age^2 - b \cdot age + c[\text{yr}^{-1}].$$

Develop a model, giving how many rabbits there are as a function of time, as
well as the age distribution of the rabbits. Calculate the number of rabbits after five
years by solving the discretized problem with time step 1 yr, $a = 0.015$, $b = 0.1$,
and $c = 0.3$.

3.13 Formulate a population balance for the number of crystals with "characteristic"
dimension (size), assuming:
- the crystallizer is operating at steady conditions and is well mixed (tank volume =
  V);
- the flow rate through the reactor is constant ($= Q$) and the inflow contains no
  crystals;
- the growth rate is v (m s^{-1});
- agglomeration and breakage of crystals can be neglected.

3.14 Use the method of population balances to develop expressions for the RTDs in the
following reactor systems:
(a) ideal tank reactor with volume V, constant flow rate q, and no chemical
reaction;
(b) as in (a) but in series with an ideal tube reactor with volume V_1;
(c) two ideal tank reactors in series (both with volume V).

4 Empirical model building

Many phenomena in engineering are very complex and we do not have sufficient knowledge at the moment to develop a model from first principles; instead, we have to rely on empirical correlations. Today most process development is done using empirical or semi-empirical models. These models are usually accurate and very useful. The drawback is that they are only valid for specific equipment within an experimental domain where the parameters are determined.

In developing a correlation, we need first to identify all the variables that may have an influence on it. There are different approaches to finding important variables. One approach is to formulate the governing equations even if we do not have sufficient knowledge or computer resources to solve the equations. These equations and the corresponding boundary conditions provide information about which variables are important in formulating an empirical correlation. A second approach, used by experienced engineers, is to list all variables that are believed to be important. The final correlation is then obtained by experimenting and model fitting using experimental design to obtain reliable results and to minimize correlations between the parameters in the model.

The recommended approach to formulating an empirical correlation is to use dimensionless numbers for describing both the dependent and the independent variables. Using dimensionless variables decreases the degrees of freedom, i.e. it decreases the number of variables and parameters in the fitting. All terms in an equation must have the same dimension, and a system that contains n variables involving m dimensions can be studied by varying the $n - m$ dimensionless variables. An additional advantage of using dimensionless variables is that correlations written in dimensionless variables are also easier to read and cause less confusion when used.

4.1 Dimensional systems

A dimensional system consists of the selected base dimensions and the dimensions of all involved variables. Table 4.1 contains a useful set of base dimensions; note that this set has the potential for additional variables for the sake of convenience, e.g. the enthalpy (in joules) is added despite the fact that it has the dimension $J = kg\ m^2\ s^{-2}$ and can be rewritten as a function of mass, length, and time. In formulating the model, we can remove one of the dimensions in enthalpy, i.e. mass, length, or time, and replace it with enthalpy to keep the base dimension at a minimum. In addition to the base dimensions

Table 4.1. A set of base dimensions

Variable	Dimension	SI units
Length	**L**	m
Mass	**M**	kg
Time	**t**	s
Temperature	**T**	K
Enthalpy	**H**	J
Amount of species	**N**	mole

Table 4.2. Dimensions of variables

Measured variable	Dimension	SI units
Force	$\mathbf{MLt^{-2}}$	$kg\ m\ s^{-2} = N$
Pressure	$\mathbf{ML^{-1}t^{-2}}$	$kg\ m^{-1}\ s^{-2} = Pa$
Power	$\mathbf{ML^2t^{-3}}$ or $\mathbf{Ht^{-1}}$	$kg\ m^2\ s^{-3} = J\ s^{-1} = W$
Density	$\mathbf{ML^{-3}}$	$kg\ m^{-3}$
Surface tension	$\mathbf{Mt^{-2}}$	$kg\ s^{-2} = N\ m^{-1}$

in Table 4.1, there are several more variables that can be selected as base dimensions if required, e.g. electric current (ampere) and magnetic field (gauss).

The dimensions of all variables can then be written as functions of the base variables in Table 4.1, as seen in Table 4.2. Depending on the selection of base dimensions, the dimension can be written in different forms.

4.2 Dimensionless equations

All models we use in chemical engineering can be formulated in the dimensionless form. The fundamental equations that describe the most phenomena in chemical engineering are formulated in Equations (4.1)–(4.5), and the involved variables are listed in Table 4.3. The characteristic dimensions are mass (M), length (L), time (t), enthalpy (H), temperature (T), and moles (N).

The momentum equation, Equation (4.1), together with the continuity equation, Equation (4.2), describe the flow, and with the heat balance, Equation (4.4), and the species balance, Equation (4.5), a complete system is described. However, this is a simplified example because most source terms are not included.

Momentum balance

$$\rho \left[\frac{\partial \mathbf{v}}{\partial t} + \mathbf{v} \cdot \nabla \mathbf{v} \right] = \mu \nabla^2 \mathbf{v} + \rho G - \nabla p \left[\frac{kg}{m^2\ s^2} \right]. \qquad (4.1)$$

Equation of continuity

$$\frac{\partial \rho}{\partial t} = -[\nabla(\rho \mathbf{v})] \left[\frac{kg}{m^3\ s} \right]. \qquad (4.2)$$

Table 4.3. Variables involved in Equations (4.1)–(4.5)

Variable	Physical property	Characteristic dimension	SI units
k	thermal conductivity	$\mathbf{HL^{-1}T^{-1}t^{-1}}$	$J\,m^{-1}\,K^{-1}\,s^{-1}$
T	temperature	$\mathbf{T}$	K
U	velocity	$\mathbf{Lt^{-1}}$	$m\,s^{-1}$
L	length	$\mathbf{L}$	m
ρ	density	$\mathbf{ML^{-3}}$	$kg\,m^{-3}$
μ	viscosity	$\mathbf{ML^{-1}t^{-1}}$	$kg\,m^{-1}\,s^{-1}$
c_p	specific heat	$\mathbf{HM^{-1}T^{-1}}$	$J\,kg^{-1}\,K^{-1}$
C	concentration	$\mathbf{NL^{-3}}$	$mol\,m^{-3}$
D	diffusivity	$\mathbf{L^2t^{-1}}$	$m^2\,s^{-1}$

As pressure is not an explicit variable in the momentum equation, the continuity equation is often rewritten by combining momentum balance and the continuity equation. For constant viscosity and density, the continuity equation can be replaced by the Poisson equation:

$$\nabla^2 p = -\nabla \cdot \rho \mathbf{v} \cdot \nabla \mathbf{v} \left[\frac{Pa}{m^2} = \frac{kg}{m^3\,s^2} \right]. \tag{4.3}$$

Heat balance (without source terms)

$$\rho c_p \left[\frac{\partial T}{\partial t} + \mathbf{v} \cdot \nabla T \right] = \nabla \cdot k \nabla T \left[\frac{J}{m^3\,s} \right]. \tag{4.4}$$

Species balance (without source terms)

$$\frac{\partial C_i}{\partial t} + \mathbf{v} \cdot \nabla C_i = \nabla \cdot D \nabla C_i \left[\frac{mol}{m^3\,s} \right]. \tag{4.5}$$

Rewriting the governing equations in the dimensionless form simplifies the estimation of the behavior of processes on different scales. Since all terms in the equations have the same dimensions, we can obtain a dimensionless equation by dividing the terms in the equations by any of the other terms. However, dividing by the term that we assume is the most important will simplify the analysis. In many cases, the convective term is the dominating term. The tables in Appendix B show how the dimensionless variables are formed from momentum, heat, and mass balances.

All variables must be scaled with a variable characteristic of the process. The characteristic velocity U may be the average inlet velocity. The characteristic length L may be the tube diameter, the reactor length, or the catalyst particle diameter, depending on the processes involved. Time may be scaled with residence time or the time constant for diffusion.

Throughout the book we will be using ˆ to denote dimensionless variables, and we will obtain

$$\hat{x} = \frac{x}{L} \quad \text{and} \quad \hat{t} = t\frac{U}{L} \quad \text{or} \quad \hat{t} = t\frac{D}{L^2}. \tag{4.6}$$

Concentration is usually scaled with the total or inlet concentrations. The characteristic temperature may be inlet temperature, but it is often scaled with the minimum and maximum temperatures to obtain a dimensionless variable between 0 and 1:

$$\hat{C} = \frac{C}{C_{TOT}} \quad \text{and} \quad \hat{T} = \frac{T - T_{\min}}{T_{\max} - T_{\min}} = \frac{T - T_{\min}}{\Delta T}. \tag{4.7}$$

Gravity is important for buoyancy, and the maximal force caused by gravity due to the difference in density is a suitable characteristic variable:

$$\hat{G} = \frac{G}{\Delta \rho g}. \tag{4.8}$$

The additional dimensionless dependent variables are formed from the characteristic velocity and the pressure difference over the system:

$$\hat{v}(\hat{x}, \hat{t}) = \frac{v(x, t)}{U}, \qquad \hat{p}(\hat{x}, \hat{t}) = \frac{p(x, t)}{\Delta p}. \tag{4.9}$$

The Reynolds, Euler, Froude, and Péclet numbers,

$$Re = \frac{\rho U L}{\mu}, \quad Eu = \frac{\Delta p}{\rho U^2}, \quad Fr = \frac{g L \Delta \rho}{\rho U^2}, \quad Pe_H = \frac{U L}{\alpha}, \quad Pe_D = \frac{U L}{D},$$

are dimensionless numbers that are most often used in formulating the transport equations, Equations (4.1)–(4.5), in the dimensionless form. The Péclet number for heat conduction, $Pe_H = U L / \alpha$, with thermal diffusion $\alpha = k / \rho c_P$, has a corresponding Péclet number for diffusion, $Pe_D = U L / D$. It is often convenient to separate the dimensionless variables into variables that describe the properties of a flow and a fluid. The Prandtl and Schmidt numbers are useful dimensionless variables that describe the fluid properties heat conduction and diffusion, respectively:

$$Pr = \frac{c_p \mu}{k} \quad \text{and} \quad Sc = \frac{\mu}{\rho D}. \tag{4.10}$$

Note that the Péclet number can also be formulated as a product of the Reynolds number, i.e.

$$Pe_H = Re \cdot Pr \quad \text{and} \quad Pe_D = Re \cdot Sc. \tag{4.11}$$

By multiplying Equations (4.1)–(4.5) by L/U and substituting the dependent variables with corresponding dimensionless variables, we obtain the following dimensionless equations.

Momentum balance

$$\frac{\partial \hat{v}}{\partial \hat{t}} + \hat{v} \cdot \nabla \hat{v} = \frac{1}{Re} \nabla^2 \hat{v} + \frac{1}{Fr} \hat{G} - \frac{1}{Eu} \nabla \hat{p}. \tag{4.12}$$

Equation of continuity

$$\nabla^2 \hat{p} = -\frac{1}{Eu} \nabla \cdot \hat{v} \cdot \nabla \hat{v}. \tag{4.13}$$

Heat balance

$$\frac{\partial \hat{T}}{\partial \hat{t}} + \hat{\mathbf{v}} \cdot \nabla \hat{T} = \frac{1}{Pe_H} \nabla^2 \hat{T}. \tag{4.14}$$

Species balance

$$\frac{\partial \hat{C}}{\partial \hat{t}} + \hat{\mathbf{v}} \cdot \nabla \hat{C} = \frac{1}{Pe_D} \nabla^2 \hat{C}. \tag{4.15}$$

It is not possible to solve these equations with unknown time-dependent boundary conditions. However, we can use the equations to identify the important variables that will affect pressure drop, heat, or mass transfer. The velocity, pressure, temperature, and concentration described with these models should be completely described by the dimensionless variables Re, Fr, Pe_H, Pe_D, and Eu. Keeping all these dimensionless numbers constant will also give the same solution expressed in the dimensionless form. Equations (4.12) and (4.13) contain Re, Fr, and Eu, and we can find correlations for pressure drop by empirically determining the function

$$f(Re, Fr, Eu) = \text{constant}; \tag{4.16}$$

we can conclude that Euler is only a function of Reynolds and Froude. We know the correlation for the laminar flow in a horizontal pipe, and, for constant density, i.e. $Fr = 0$, we have the Hagen–Poiseuille equation,

$$\frac{\Delta p}{L} = \frac{32 \mu U}{d^2}, \tag{4.17}$$

that, in the dimensionless form using $\Delta p = Eu \cdot \rho U^2$, becomes

$$\frac{Eu \cdot \rho U^2}{L} = \frac{32 \mu U}{d^2} \quad \text{or} \quad Eu = \frac{32 \mu U L}{\rho U^2 d^2} \Rightarrow Eu = \frac{32}{Re} \frac{L}{d}. \tag{4.18}$$

A corresponding correlation for turbulent flow is

$$Eu = 0.158 Re^{-0.25} \frac{L}{d}. \tag{4.19}$$

Example 4.1 Fluid–particle heat transfer

Find a suitable dimensionless model for heat transfer between a fluid and a spherical body, as seen in Figure 4.1. Heat transfer to a particle in a turbulent flow is a complex phenomenon involving convection and conduction. The objective is to develop an empirical correlation that can describe the time-averaged heat transfer coefficient h.

The characteristic variables must be identified in order to formulate a dimensionless model. In this case, our aim is to formulate the average heat transfer; the interesting dependent variable is heat flux, and the independent variables are temperature, velocity, and size. If we are testing different materials, then the thermal conductivity, specific heat, density, and viscosity are also independent variables. Here we will use the characteristic length L as the diameter of the sphere and the characteristic velocity U as the velocity far from the body. The characteristic temperature is set as the difference

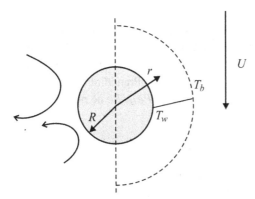

Figure 4.1. Heat transfer to a spherical particle.

between the bulk temperature and the surface temperature, i.e. $\Delta T = T_b - T_w$. Time is not relevant here, but is discussed in Section 5.1.5. The governing equations and the characteristic dimensions are written in the following using particle diameter as the characteristic length $L = d$.

The film theory describes heat transfer to the surface. The heat flux modeled using the heat transfer coefficient is balanced with the flux into the solid sphere:

$$q = h(T_b - T_w) = k \left. \frac{dT}{dr} \right|_{r=R} \left[\frac{\text{J}}{\text{m}^2 \, \text{s}} \right]. \tag{4.20}$$

Defining $\hat{T} = (T - T_b)/\Delta T$ and $\hat{r} = r/(d/2)$ with $Nu = h \cdot d/k$, we obtain, after dividing by ΔT and h, the dimensionless equation for heat transfer to the surface.

Heat transfer at the particle surface

$$1 = \frac{1}{Nu} \left. \frac{d\hat{T}}{d\hat{r}} \right|_{\hat{r}=1}. \tag{4.21}$$

Removing the transient part from Equations (4.12)–(4.14) gives the remaining equations.

Momentum balance

$$\hat{\mathbf{v}} \cdot \nabla \hat{\mathbf{v}} = \frac{1}{Re} \nabla^2 \hat{\mathbf{v}} + \frac{1}{Fr} \hat{G} - \frac{1}{Eu} \nabla \hat{p}. \tag{4.22}$$

Equation of continuity

$$\nabla^2 \hat{p} = -\frac{1}{Eu} \nabla \hat{\mathbf{v}} \cdot \nabla \hat{\mathbf{v}}. \tag{4.23}$$

Heat balance

$$\hat{\mathbf{v}} \cdot \nabla \hat{T} = \frac{1}{Pe_H} \nabla^2 \hat{T}. \tag{4.24}$$

Velocity and pressure are determined from Equations (4.22)–(4.23), i.e. $Eu = f(Re, Fr)$. For simplicity, we can assume that there are no density differences, and the term containing the Froude number can be neglected. Only Nu, Re, Eu, and Pe_{II} remain, and as $Eu = f(Re)$ the Nusselt number Nu must be a function of Re and Pe_H. However, in most empirical correlations, the Prandtl number, Pr, is preferred instead of the Péclet number, Pe_H, but as $Pr = Pe_H/Re$ no new dimensionless variables are introduced:

$$Nu = b_0 + b_1 Re^{\alpha_1} Pr^{\beta_1} + b_2 Re^{\alpha_2} Pr^{\beta_2} + \cdots. \qquad (4.25)$$

The parameters b_i, α_i, and β_i are determined from experimental data. One common correlation is the Frössling/Ranz–Marshall correlation,

$$Nu = 2 + 0.6 Re^{1/2} Pr^{1/3}. \qquad (4.26)$$

Using dimensionless variables allows for the development of scale-independent design equations. Note that the species balance equation is almost identical to the heat balance, and that, by replacing the Pr number in Equation (4.25) with the Schmidt number Sc, we obtain a similar correlation for mass transfer as for heat transfer. In principle they describe similar phenomena, and, if we replace heat conduction k with heat diffusion $\alpha = k/\rho c_p$ containing the same dimension as diffusivity (m^2 s^{-1}), we obtain the same expression.

Many different dimensionless numbers can be obtained by simply dividing the terms in any dimensional consistent equation, as seen in Appendix B. The most commonly used dimensionless numbers are listed in Table B.4.

4.3   Empirical models

The dimensionless form is also suitable for developing models or correlations that have a less theoretical basis and cannot be deduced from simple balance equations. The advantage of using dimensionless variables is that, if there are any n variables containing m primary dimensions, the correlation can be formulated using $n - m$ dimensionless groups. However, we need experience or some preliminary experiments to identify the variables that are important.

To illustrate this method, we will use Example 4.1 with heat transfer to a particle or a wall. Fluid solid heat transfer can be modeled as in Equation (4.20) as follows:

$$q = h(T_b - T_w) = k \left. \frac{dT}{dr} \right|_{r=R}. \qquad (4.20)$$

The objective is to develop a model that can predict the heat transfer coefficient h or the Nusselt number. Equation (4.20) says nothing about the heat transfer coefficient h, and we need to conduct experiments to develop a model of heat flux that includes convection, heat conduction, viscosity, heat capacity, etc., so that we can predict the

Table 4.4. Basic dimensions as functions of the variables in the recurring set of variables

Variables	Dimension	SI units
Length	$\mathbf{L} = l$	m
Mass	$\mathbf{M} = \rho l^3$	kg
Time	$\mathbf{t} = \rho l^2 \mu^{-1}$	s
Temperature	$\mathbf{T} = \Delta T$	K
Enthalpy	$\mathbf{H} = k l^3 \rho \mu^{-1} \Delta T$	J

Nusselt number. In formulating a general correlation, we can list all variables that we believe will affect heat transfer:

$$q = f(k, \Delta T, U, l, \rho, \mu, c_p, \beta g). \tag{4.27}$$

These n variables have dimensions in length L (m), mass M (kg), time t (s), temperature T (K), and enthalpy H (J). The variables with their dimensions are listed in Table 4.3. In contrast to Example 4.1, we allow free convection, and an additional variable βg describes the effect of gravity on free convection. The thermal expansion coefficient β describes how fluid density changes with temperature, e.g. for an ideal gas $\beta = 1/T$. A correlation for heat transfer should include these variables in a dimensional consistent form.

The dimensionless variables can be found by selecting m variables as a *recurring set*. These variables should contain the different variables that form a complete set of dimensions, i.e. it should not be possible to formulate a dimensionless variable by combining the variables in the recurring set. Here we will select $l, \rho, \mu, \Delta T, k$ as the recurring set. They are listed with their dimensions in Table 4.4. The selection of variables is not critical, and other combinations of variables containing length, mass, time, temperature, and enthalpy that could form the recurring set of variables are possible.

Selecting $l, \rho, \mu, \Delta T, k$ as the recurring set leaves the $n - m$ variables $q, U, \beta g, c_p$ as the *non-recurring* variables. Using the dimensions in Table 4.4, the dimensional variables are then identified as functions of the recurring variables. The non-recurring variables q [J m^{-2} s^{-1}], U [m s^{-1}], c_p [J kg K^{-1}], and βg [m s^{-2} K^{-1}] can be reformulated into dimensionless variables, e.g. q [J m^{-2} s^{-1}] can be made dimensionless by dividing by enthalpy H (J) and multiplying by length (L^2) and time (t). After replacing H, L, and t by the variables in Table 4.4 and q by $h\Delta T$, we obtain

$$\Pi_1 = \frac{q\mathbf{L}^2\mathbf{t}}{\mathbf{H}} = \frac{q(l^2)(\rho l^2/\mu)}{(kl^3\rho\Delta T/\mu)} = \frac{ql}{k\Delta T}\frac{h}{h} = \frac{hl}{k} = Nu,$$

$$\Pi_2 = \frac{U}{\mathbf{Lt}} = \frac{U\rho l}{\mu} = Re,$$

$$\Pi_3 = \frac{c_p\mathbf{MT}}{\mathbf{H}} = \frac{c_p\mu}{k} = Pr,$$

$$\Pi_4 = \frac{\beta g\mathbf{t}^2\mathbf{T}}{\mathbf{L}} = \frac{\beta g\Delta T\rho^2 l^3}{\mu^2} = Gr. \tag{4.28}$$

The use of Π as the symbol for dimensionless variables was introduced by Buckingham, and this method has been called Buckingham's Π theorem since then. This is a very general and useful method, but it will not guarantee the most physical meaningful correlation.

According to Equation (4.27), heat flux should be a function of the remaining variables, which allows us to write the dimensionless heat flux as a function of the remaining dimensionless numbers:

$$\Pi_1 = f(\Pi_2, \Pi_3, \Pi_4) \quad \text{or} \quad Nu = f(Re, Pr, Gr). \tag{4.29}$$

The dimensionless analysis tells us nothing about the correlation; it only reveals which variables are involved. The final correlations can be formulated in many different ways. An empirical model for heat transfer may be formulated as

$$Nu = b_0 + b_1 Re^{\alpha_1} Pr^{\beta_1} Gr^{\gamma_1} + b_2 Re^{\alpha_2} Pr^{\beta_2} Gr^{\gamma_2} + \cdots, \tag{4.30}$$

and the parameters b_i, α_i, β_i, and γ_i can be fitted from measurements. The dimensionless number Gr includes the effect of buoyancy. For heat transfer to a sphere in Example 4.1, with no buoyancy, the following empirical model has been found, as shown previously:

$$Nu = 2 + 0.6 Re^{1/2} Pr^{1/3}; \tag{4.26}$$

for free convection on a vertical wall,

$$Nu = 0.68 + \frac{0.67 Ra^{1/4}}{[1 + (0.492/Pr)^{9/16}]^{4/9}}, \tag{4.31}$$

where $Ra = Gr \cdot Pr$.

4.4   Scaling up

Scaling up equipment is a general problem in chemical engineering. If we were running all processes at thermodynamic equilibrium, there would be no scaling-up problem and the final composition could be calculated from thermodynamic data. However, this solution would require infinitely large equipment, and the thermodynamic equilibrium may not be the composition we want. In industrial production we aim for maximum production in the plant whilst keeping the quality of the product high. In scaling up we cannot maintain the same conditions on all scales, e.g. surface/volume will decrease with increasing size. Thus we introduce limitations on mass and heat transfer, and cannot expect the same results on the industrial production scale as on the laboratory scale.

Before we build an industrial-scale process, we need to use the laboratory- and pilot-plant-scale equipment to develop an understanding of how the industrial-scale production plant will work. The models we develop on smaller scales must be scalable and applicable to the large scale.

In general, we can separate the three different cases: scaling up, scaling down, and scaling out. In scaling up, we aim for building new equipment and will use experiments on a small scale to develop models that are scale independent and can be used for designing the full-scale equipment. In scaling down, we already have the large-scale equipment that we want to optimize for a new process. In such a case, we want to know how to design and run a pilot plant to mimic the behavior of a large-scale plant in order to find the optimal conditions for it. In scaling out, we build a large-scale plant by putting the small-scale equipment in parallel. We then expect to obtain the same conditions in the large scale as in the small scale. This can be done by, e.g., using small tubular reactors in parallel instead of building a large tubular reactor. The scaling-up problem is then limited to obtaining the same flow rate in all tubes and sufficient heat control of the tube bundle to keep the temperature equal.

The fundamental principle in scaling models is to keep all dimensionless variables constant. The dimensionless equations, e.g. Equations (4.12)–(4.15), are scale independent, and keeping Re, Pr, Eu, and Fr numbers constant results in dimensionless variables $\hat{\mathbf{v}}$, $\hat{\mathbf{G}}$, $\hat{p}$, $\hat{T}$, and $\hat{C}$ versus $\hat{t}$ that is the same at all scales. In the simplest case, scaling with a constant Re, i.e. an increase in length scale by a factor of 10, requires that the characteristic velocity decreases by a factor of 10. As a result, assuming no density difference, i.e. $Fr = 0$, Eu will be constant and the pressure drop is scaled with ρU^2. Keeping Re and Pr constant should also result in the same dimensionless heat transfer coefficient, i.e. the Nusselt number, as obtained in the dimensionless correlations in Example 4.1.

For free convection problems, the Froude number must also be kept constant. Since we cannot change gravity, the density change with temperature and $\Delta\rho/\rho$ must be increased. This might be very difficult while keeping Re and Pr constant.

More complex phenomena require keeping more dimensionless variables constant. A more complete heat flux balance will contain more terms, and consequently more dimensionless variables must be kept constant:

$$\rho c_p \mathbf{v} \cdot \nabla T = \nabla \cdot k \nabla T + r_A \Delta H + ha \Delta T + e\sigma T^3 \Delta T. \qquad (4.32)$$

convection   conduction   reaction   phase transfer   radiation

Dividing by $\rho c_p U \Delta T L$, we obtain

$$\hat{\mathbf{v}} \cdot \nabla \hat{T} = \frac{1}{Pe} \nabla^2 \hat{T} + \frac{1}{Da_{III}} \hat{r}_A + \frac{Nu}{RePr} \hat{T} + \frac{e\sigma T^3 L}{\rho c_p U} \hat{T}. \qquad (4.33)$$

In addition to the dimensionless variables for flow, Re, Fr, and Eu, we also need to keep Pe, Nu, Da_{III}, and $e\sigma T^3 L/\rho c_p U$ constant to obtain the same dimensionless predicted conditions, i.e. $Re = \rho U L/\mu$, $Pr = \mu/\rho\alpha$, $Eu = \Delta p/\rho U^2$, $Fr = gL\Delta\rho/\rho U^2$, $Pe = \rho c_p U L/k$, $Nu = hL/k$, $Da_{III} = r_A L\Delta H/\rho c_p U \Delta T$, and $e\sigma T^3 L/\rho c_p U$ should be kept constant.

We quickly reach a level of complexity that makes it impossible to keep all dimensionless variables constant. Using correlations developed on a small scale to predict the behavior on the large scale may not be a good solution, because empirical correlations are only correlations and as such are only valid within the range of experiments that form the basis for the correlations. As not all the dimensionless variables could be kept constant on all scales in our example, no experimental data from the small scale are

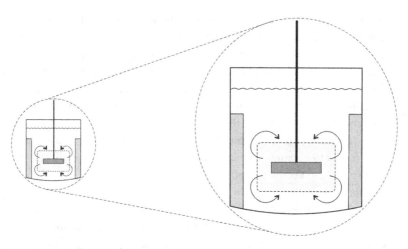

Figure 4.2. Scaling up of stirred tank reactors.

valid for the large scale. Even if the correlations are written in the dimensionless form, this limitation must be kept in mind.

The next part of the procedure is to identify the rate-determining step, and to determine which conditions should be scale similar. All the dimensionless variables may not be important. There are no general methods for finding the rate-determining steps, and a genuine understanding of the actual processes is needed. Detailed knowledge of specific processes is beyond the scope of this book. However, the example below will illustrate this approach.

Example 4.2 Mixing and reaction in a stirred tank reactor

Assume that the chemical reaction $A + B \rightarrow$ products is scaled up from a 10 l pilot plant to 10 m^3 full-scale production unit. Temperature control is usually a minor problem in stirred tank reactors; instead the critical property is the rate of mixing. The characteristic reaction time can be estimated from the kinetics and the concentration as follows:

$$r = k(T) \cdot f(C_A, C_B) \left[\frac{mol}{m^3 \, s} \right] \quad \text{and} \quad \tau_{react} = \frac{C_{limit}}{r} \, [s],$$

where C_{limit} is the concentration of the limiting reactant.

The rate-limiting steps may be the kinetics, but the macro- or micromixing can also be rate limiting. A general approach to this problem requires that a large number of dimensionless variables be kept constant. If the rate-determining step(s) can be identified, the scale-up process is easier.

In a very slow chemical reaction, the time for mixing is much faster than the reaction rate. The process is only dependent on species concentration and temperature. Keeping these constant during scale-up will result in the same production per volume reactor. However, the surface area per volume will decrease with size, and additional cooling/heating coils may be required.

In a slow chemical reaction, the flow circulation time in the reactor is on the same time scale as the chemical reaction time, which will lead to variations in concentration within the reactor. Scaling up with constant circulation times would be a proper scaling-up rule. However, the flow must be turbulent in both cases, or the mixing time will increase by several orders of magnitude.

The pumping capacity of an impeller is proportional to the impeller tip speed $N \cdot \pi \cdot d$ and the swept area of the impeller tip $\pi \cdot d \cdot h$. Here, N is the impeller speed, d is the impeller diameter, and h is the height of the impeller blades. Assuming that the impeller blades scale with the reactor size, D, we can assume that d/D and h/D are constant on all scales, and consequently the pumping rate is given by

$$Q \propto N\pi d \cdot \pi d \cdot h \propto ND^3 \ [\text{m}^3 \ \text{s}^{-1}], \tag{4.34}$$

and, following the scaling-up rule with a constant pumping rate per volume reactor, circulation time is given by

$$\tau_{circ} \propto \left(\frac{D^3}{ND^3}\right)_{large} = \left(\frac{D^3}{ND^3}\right)_{small} \quad \text{or} \quad \tau_{circ} \propto \left(\frac{1}{N}\right)_{large} = \left(\frac{1}{N}\right)_{small}. \tag{4.35}$$

This scaling rule indicates that the impeller speed should be kept constant. However, the energy input from the impeller scales with $N^3 D^5$, as shown in Practice problem 4.3, and at constant impeller speed the required power input scales as $(D_{large}/D_{small})^5$. An increase by a factor of 10 in linear dimension will require an increase by a factor of 10^5 in power input. This is usually not possible to obtain, and a larger variation in concentration will occur on the large scale.

Fast chemical reactions are affected by the mixing on the smallest scale. The turbulent properties together with the kinetics determine the outcome. The turbulence is generated by the impeller, but turbulence is also dissipated much faster in the impeller region. This will lead to two very different regions in the reactor: the impeller region with very fast mixing and reaction, and the remaining volume with very slow mixing and reaction.

The mixing time for fast reactions scales with the lifetime of the turbulent eddies k/ε or, for more viscous, fluids as $\sqrt{v/\varepsilon}$, where k is the turbulent kinetic energy, ε is the rate of dissipation, and v is the kinematic viscosity. The kinetic energy scales as the impeller tip speed squared, i.e. $k \propto (ND)^2$, and the dissipation scales as power input per volume, $\varepsilon \propto P/V \propto N^3 D^5/D^3 = N^3 D^2$. The micromixing time is then estimated as

$$\tau_{micro} \propto \frac{k}{\varepsilon} \propto \frac{N^2 D^2}{N^3 D^2} = \frac{1}{N} \quad \text{or} \quad \tau_{micro} \propto \sqrt{\frac{v}{\varepsilon}} \propto N^{-3/2} D^{-1}. \tag{4.36}$$

Keeping k/ε constant requires the same impeller speed on all scales, and consequently we obtain the same problem with power input as for constant circulation time. Constant ε requires that when size D increases, the impeller speed should decrease as

$$N_{large} = N_{small} \left(\frac{D_{small}}{D_{large}}\right)^{2/3}. \tag{4.37}$$

Here we have assumed that the turbulent kinetic energy and the dissipation rate are the same throughout the whole reactor. However, stirred tank reactors are very heterogeneous, with much higher kinetic energy and a higher dissipation rate in the impeller region than in the rest of the tank. This makes stirred tank reactors very difficult to scale-up. Reactors containing packed beds, metal foams, or static mixers are much easier to scale-up since the flow, mass transfer, and reactions are much more similar on the small and large scales.

4.5 Practice problems

4.1 The Ergun equation for pressure drop in porous beds is given by

$$\frac{\Delta P}{L} = -A\mu U - B\left(\frac{1}{2}\rho U^2\right),\tag{4.38}$$

with

$$A = \frac{150(1-\alpha)^2}{d_p^2\alpha^3} \quad \text{and} \quad B = \frac{1.75(1-\alpha)}{d_p\alpha^3},$$

where α is the void fraction.

Show that the pressure drop can be written in the form

$$Eu = b_0 + b_1 Re^{\beta_1} + b_2 Re^{\beta_2}.$$

4.2 Derive an expression for the Sherwood number as a function of Re and Sc using Equations (4.12), (4.13), and (4.15), and suitable boundary conditions.

4.3 The power input per unit volume in a stirred tank reactor depends on liquid density, the impeller diameter, and speed, i.e. $P = f(\rho, D, N)$. Write a dimensionless relation between power input and these variables. Assume that the impeller and tank size have a constant ratio, and only one characteristic length, the impeller diameter D, is needed.

4.4 In boiling, the heat transfer coefficient h [W m^{-2} K^{-1}] depends on the heat flux q [W m^{-2}], thermal conductivity k [W m^{-1} K^{-1}], latent heat of vaporization λ [J kg^{-1}], viscosity μ [kg m^{-1} s^{-1}], surface tension σ [J m^{-2}], pressure, P [N m^{-2}], bubble diameter d [m], liquid density ρ [kg m^{-3}], and the density difference between liquid and vapor $\Delta\rho$ [kg m^{-3}]:

$$h = f(q, k, \lambda, \mu, \sigma, P, d, \rho, \Delta\rho).$$

Formulate a model for the heat transfer coefficient (Nusselt number) with the minimum number of dimensionless variables.

5 Strategies for simplifying mathematical models

A mathematical model can never give an exact description of the real world, and the basic concept in all engineering modeling is, "All models are wrong – some models are useful." Reformulating or simplifying the models is not tampering with the truth. You are always allowed to change the models, as long as the results are within an acceptable range. It is the objective of the modeling that determines the required accuracy: Is it a conceptual study limited to order of magnitude estimations? Or is it design modeling in which you will add 10–25% to the required size in order to allow for inaccuracies in the models and future increase in production? Or is it an academic research work that you will publish with as accurate simulations as possible?

A simulation may contain both errors and uncertainties. An *error* is defined as a recognizable deficiency that is not due to lack of knowledge, and an *uncertainty* is a potential deficiency that is due to lack of knowledge. All simulations must be validated and verified in order to avoid errors and uncertainties. Validation and verification are two important concepts in dealing with errors and uncertainties. *Validation* means making sure that the model describes the real world correctly, and *verification* is a procedure to ensure that the model has been solved in a correct way.

Poor simulation results are due to many different reasons:

- errors in formulating the problem;
- inaccurate models;
- inaccurate data on the physical properties;
- numerical errors due to the choice of numerical methods;
- lack of convergence in the simulations.

It is always recommended that you do simple by-hand calculations to define the properties of the system prior to formulating the final model. The objective of this chapter is to present strategies for simplifying given mathematical models and for reducing the models already in the formulation.

Many models in chemical engineering are derived from four basic balance equations, as seen in Chapter 4, but now we have also included the source terms for chemical reactions and heat formation.

Material balance

$$\frac{\partial C_A}{\partial t} + \mathbf{v} \cdot \nabla C_A = \nabla \cdot D_A \nabla C_A - R_A. \tag{5.1}$$

Heat balance

$$\rho c_p \left[\frac{\partial T}{\partial t} + \mathbf{v} \cdot \nabla T \right] = \nabla \cdot k \nabla T + Q. \tag{5.2}$$

Momentum balance

$$\rho \left[\frac{\partial \mathbf{v}}{\partial t} + \mathbf{v} \cdot \nabla \mathbf{v} \right] = \mu \nabla^2 \mathbf{v} + \rho G - \nabla p. \tag{5.3}$$

Equation of continuity

$$\frac{\partial \rho}{\partial t} = -[\nabla(\rho \mathbf{v})], \tag{5.4}$$

with suitable initial and boundary conditions.

There are general programs that can solve these equations for simple systems, e.g. computational fluid dynamics, CFD. However, the equations can easily become very difficult if they include stiff non-linear equations and complex geometry that call for a structured approach.

5.1   Reducing mathematical models

Equations (5.1)–(5.4) describe many systems in chemical engineering. But the equations do not contain the true nature of the real world; they are merely convenient mathematical approximations of our models of the real world.

These complex mathematical models are in many cases difficult to solve. A very good work station is necessary for solving 3D flows in a reactor with an exothermic chemical reaction. If we want to solve the equations for a heterogeneous system, a computer cluster is required that can give a solution within a reasonable time. Yet in many cases the equations can be reduced without losing too much accuracy. In some cases, an even better numerical precision can be reached. The following list covers some of the most commonly used methods for reducing the computational efforts for solving mathematical equations.

- decoupling equations;
- reducing the number of independent variables;
- averaging over time or space, "lumping";
- simplifying geometry;
- separating equations into steady state and transient;
- linearizing;
- solving for limiting cases;
- neglecting terms;
- changing boundary conditions.

5.1.1   Decoupling equations

Decoupling equations is one of the most important tools for reducing complex mathematical models. A model with many dependent variables, e.g. different velocities, concentrations of chemical substances, and temperature, can be simplified substantially if we look for variables that are not dependent on the others. In Equations (5.1)–(5.4), the coupling occurs mainly through the velocity in all the equations and through temperature. The reaction rate in Equation (5.1) depends on temperature, and the temperature in Equation (5.2) depends on the reaction rate through the heat of the reaction. Viscosity and density of the fluid also depend on temperature and chemical composition, and as a consequence all equations must be solved simultaneously. On the contrary, if the heat of the reaction is low, the temperature variation will be small, the fluid density and viscosity can be assumed constant, and the flow may be calculated in advance. In such a case, the chemical reactions can then be added to a precalculated velocity distribution.

Coupling may also occur via boundary conditions, e.g. the reaction rate in a catalyst pellet depends on the concentration and the temperature of the fluid surrounding the pellet. At steady state, when coupling between equations occurs through boundary conditions, an exact or approximate analytical solution can be calculated with boundary conditions as variables, e.g. the effectiveness factor for a catalyst particle can be formulated as an algebraic function of surface concentrations and temperature. The reaction rate in the catalyst can then be calculated using the effectiveness factor when solving the reactor model. However, this is not possible for transient problems. The transport in and out of the catalyst also depends on the accumulation within the catalyst, and the actual reaction rate depends on the previous history of the particle.

A very efficient method is to precalculate steady-state solutions and save the results in look-up tables. Interpolation in multidimensional look-up tables is most often much faster than iterative solutions of non-linear equations. It is also possible to use look-up tables for initial-value differential equations using time as one variable in the look-up table.

Decoupling equations may also be used to enhance the numerical stability of stationary problems even when the equations have strong coupling. You can start by solving, e.g., the flow field and then continue with species, reactions, and heat, and finally solve the flow field, species, and heat with updated properties for concentrations, temperature, density, and viscosity. This control of the iterations allows for a more stable approach to the final solution.

5.1.2   Reducing the number of independent variables

By reducing the number of independent variables, e.g. space or time, a partial differential equation can, in the best case, be reduced to an ordinary differential equation or even an algebraic equation, and the computational effort is reduced substantially. The most obvious approach is to use the symmetry in geometry. Spherical and cylindrical symmetry is often easy to find. This will reduce a 3D problem to a 1D or 2D problem with a significant decrease in computational time.

Geometrical symmetry requires that the boundary conditions as well as the equations are symmetrical. Problems including gravity can only be symmetrical in a cylinder with gravity in the axial direction. Reduction to a 1D or 2D problem may also change the physical appearance, e.g. bubbles do not exist in axisymmetric 2D except on the symmetry axis; they become toroid in 3D elsewhere. A toroid bubble moving in a radial direction must alter the diameter to maintain the volume, and the forces around the bubble will be unphysical.

Symmetry boundary conditions should be used with care. The outflow across a symmetry boundary is balanced by an identical inflow, which means that the net transport will be zero. Transport along the boundary is allowed, but no transport across the boundary will occur. It is not sufficient for a conclusive result that the differential equations and the boundary conditions are symmetrical; the symmetrical solution may be only one of many possible solutions.

5.1.3   Lumping

Using the average over one dimension or over a volume, i.e. lumping, may reduce the complexity of a problem. This will reduce partial differential equations to ordinary differential equations or algebraic equations. Separation processes are often modeled as equilibrium stages, as shown in Figure 2.1. Very complex flow and mass transfer conditions are lumped together into an ideal equilibrium stage with an efficiency factor to compensate for the non-ideal behavior. A second frequently used simplification is transforming a tubular reactor with axial and radial dispersion to tank reactors in series by assuming negligible variations in the radial direction and over a short axial distance. Tanks in series may also be viewed as an implicit numerical solution of a model for a tubular reactor. Implicit numerical methods are, in general, more stable than explicit methods, and the tanks-in-series model can also be chosen for numerical stability reasons.

5.1.4   Simplified geometry

Symmetry can be used to limit the volume over which the problem has to be solved. If the boundary conditions are the same on all sides, a square can be reduced to 1/8 of its surface area, as shown in Figure 5.1, and a cube can be reduced to 1/8 of its volume by modeling one corner, or even to 1/16 or 1/32 by finding more mirror planes. The boundary conditions must be the same on all external surfaces. The use of mirror planes does not reduce the number of independent variables, but it results in a smaller volume and fewer mesh points to solve.

Symmetry boundary conditions reduce the computational effort due to the requirement that all solution variables have zero normal gradient at the symmetry plane. As such, only one symmetric solution is obtained, but in reality several solutions may exist.

Periodic boundary conditions are very useful when a system has periodic behavior. The inflow at one boundary is set as the outflow from another surface, e.g. only 1/6 of a tank reactor with six impeller blades and no baffles is calculated by using the tangential outflow on one side as the inflow on the other side.

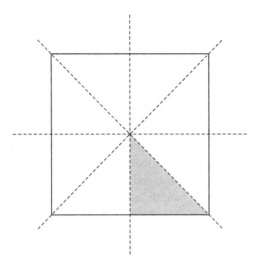

Figure 5.1. Symmetry in a square.

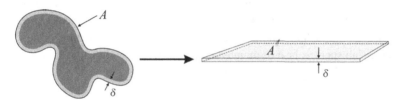

Figure 5.2. Simplified geometry. When only the outer part of a complex body is active, the body may be simplified to a slab.

In many cases, irregular-shaped particles or a size distribution of particles can be treated as spheres with an equivalent diameter that gives the same surface/volume ratio as the irregular-shaped particle. The Sauter mean diameter, usually denoted d_{32}, defined as the total volume divided by the total surface area, is a very common method used to characterize a size distribution of bubbles or drops. With this definition, the total volume can be calculated from the number of particles multiplied by the volume of a sphere with diameter d_{32}, and the total surface area can be calculated by multiplying the number of particles by the surface of the sphere.

The use of hydraulic diameter in calculating pressure drop is a commonly used concept for simplifying geometry. The hydraulic diameter for an opening is calculated by multiplying the open surface area by 4, then dividing by the wetted periphery, e.g. for a cylindrical opening $d = 4(\pi d^2/4)/(\pi d)$, and for a square duct $d = 4d^2/(4d)$.

It is not necessary to solve the complete geometry if only part of a volume is involved based on the concept that "the earth is locally flat." For example, a very fast chemical reaction in an irregular-shaped catalyst will involve only the outer surface, and therefore the catalyst may be treated as a slab with a surface A, equal to the particle surface and thickness δ, which is sufficient for containing all interesting phenomena (see Figure 5.2).

5.1.5   Steady state or transient

When different parts of a process have very different time dependencies, the fast varying variables can be treated as steady-state variables when solving for the slow variables. Similarly, the slowly varying variables can be considered constant when solving for the fast varying variables. Time constants are very convenient tools for estimating time dependence. Assuming a pseudosteady state for a complex reaction sequence, e.g. combustion involving hundreds of reactions, is a useful method for decreasing the number of ordinary differential equations that must be solved by transforming them into algebraic equations. "Pseudosteady state" means that the whole process is unsteady, but some sequences of the process change very fast, and the accumulation term in that step can be neglected. This is valid for many radicals in complex chemical reactions. The radicals react very quickly and have very low concentrations, because once they are formed they will react further.

Estimating the time constants for convection is straightforward. The residence time is estimated from length and velocity, $t = L/v$, or from volume and volumetric flow, $t = V/q$. Transport by diffusion is more difficult to estimate. Solving the unsteady transport equation for diffusion, Equation (5.5), heat conduction, or viscous transport for transport into a slab will give information about two things: the time needed to bring about a change in boundary conditions to affect the whole volume, or how far, into an infinite slab species, heat or momentum can penetrate within a given time.

Unsteady diffusion in a slab is described by

$$\frac{\partial C}{\partial t} = D \frac{\partial^2 C}{\partial z^2}, \tag{5.5}$$

which in dimensionless form is

$$\frac{\partial \hat{y}}{\partial \hat{t}} = \frac{\partial^2 \hat{y}}{\partial \hat{z}^2}, \tag{5.6}$$

with

$$\hat{x} = \frac{x}{L}, \quad \hat{y} = \frac{C}{C(\hat{z} = 0)}, \quad \hat{t} = \frac{tD}{L^2}.$$

Equation (5.6) can also describe heat transfer and slow viscous flow using

$$\hat{y} = \frac{T - T_{init}}{T(\hat{z} = 0) - T_{init}}, \quad \hat{t} = \frac{t\lambda}{\rho c_p L^2}$$

for heat and

$$\hat{y} = \frac{U}{U(\hat{z} = 0)}, \quad \hat{t} = \frac{t\mu}{\rho L^2}$$

for velocity. The boundary conditions with constant concentration on one side and no transport out on the other side,

$$\hat{y} = 1 \text{ at } \hat{z} = 0 \quad \text{and} \quad \frac{d\hat{y}}{d\hat{z}} = 0 \text{ at } \hat{z} = 1,$$

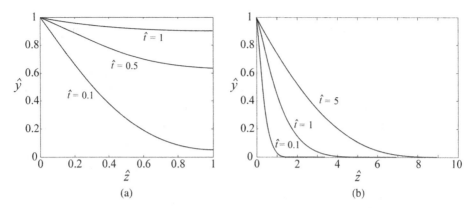

Figure 5.3. Time constants for diffusion in a slab with different boundary conditions. (a) No flux at $\hat{z} = 1$; (b) $\hat{y}(\hat{t}, \infty) = 0$.

together with the initial conditions

$$\hat{y}(0, \hat{z}) = 0,$$

describe a step change at the boundary from 0 to 1, and Equation (5.6) has the solution

$$\hat{y}(\hat{t}, \hat{z}) = \sum_{i=0}^{\infty} (-1)^i \left[\mathrm{erfc}\left(\frac{\hat{z} - 2i}{2\sqrt{\hat{t}}} \right) + \mathrm{erfc}\left(\frac{2i + 2 - \hat{z}}{2\sqrt{\hat{t}}} \right) \right]. \qquad (5.7)$$

From Figure 5.3(a) we can see that steady state at $\hat{y} = 1$ is reached when $\hat{t} \gg 1$, i.e. $\hat{t} = Dt/L^2 \gg 1$ or $t \gg L^2/D$. It is also very convenient to define dimensionless time as $\hat{t} = t/\tau$ using the time constants $\tau = L^2/D$, $\tau = \rho c_p L^2/\lambda$, and $\tau = \rho L^2/\mu$ for diffusion, heat conduction, and the viscous transport of momentum, respectively. In Figure 5.3(a), $\hat{y}$ is close to 1 when $\hat{t} = 1$, and in this case we can, with reasonable confidence, assume steady state when $t > \tau$.

Information about how far the species will penetrate as a function of time is obtained by changing the boundary condition at $\hat{z} = 1$ for Equation (5.6) to $\hat{y}(\hat{t}, \infty) = 0$. This boundary condition gives the solution

$$\hat{y}(\hat{t}, \hat{z}) = 1 - \mathrm{erf}\left(\frac{\hat{z}}{2\sqrt{\hat{t}}} \right). \qquad (5.8)$$

The result is shown in Figure 5.3(b). As $\hat{y}$ depends only on $\hat{z}/\sqrt{4\hat{t}}$, there is a linear relation between $\hat{z}$ and $\sqrt{\hat{t}}$ at constant $\hat{y}$, and putting $\hat{z}/\sqrt{\hat{t}} = 1$, i.e. $z = \sqrt{Dt}$, will yield a constant value of $\hat{y} = 0.48$, and $\hat{z}/2\sqrt{\hat{t}} = 1$ results in $\hat{y} = 0.16$. We can then estimate how far into a fluid a species has reached by diffusion after a certain time, e.g. the concentration of a species 1 mm into a fluid has reached 0.48 of the boundary condition after 1000 s with a liquid diffusivity of 10^{-9} m^{-2} s^{-1}. Using dimensionless time $\hat{t} = t/\tau$ and the time constants $\tau = \rho c_p L^2/\lambda$ or $\tau = \rho L^2/\mu$, we can calculate heat transport and the viscous transport of momentum, respectively.

Knowledge of the relative magnitude of the time constants involved in dynamic processes is often very useful in the analysis of a given problem, since this knowledge may be used to:

- reduce the complexity of mathematical models by focusing on relevant time frames;
- check the difficulty of numerical solutions due to the stiffness of the equations;
- determine whether the overall rate process is limited by a particular process, e.g. chemical reaction rates limited by kinetics or by diffusion, mixing, etc.;
- discover whether a change in the rate-determining step occurs during scale-up.

Time constants are discussed further in the case study given in Section 5.2.3.

Example 5.1 Modeling an unsteady 1D catalytic tubular reactor

The time constants in an unsteady catalytic tubular reactor may be very different for different phenomena. The concentration may change in seconds, the temperature in minutes, and the catalyst activity in weeks.

A material balance for a species and heat transported in a tubular reactor can be expressed as

$$\underbrace{\frac{\partial C}{\partial t}}_{\text{accumulation}} = \underbrace{D_{ea}\frac{\partial^2 C}{\partial x^2}}_{\text{dispersion}} - \underbrace{v\frac{\partial C}{\partial x}}_{\text{convection}} - \underbrace{k(T,t)\cdot C^n}_{\text{reaction}}, \tag{5.9}$$

$$\rho_m c_{pm}\frac{\partial T}{\partial t} = \lambda_{ea}\frac{\partial^2 T}{\partial x^2} - v\rho c_p\frac{\partial T}{\partial x} - k(T,t)(-\Delta H)\cdot C^n, \tag{5.10}$$

where ρ_m and c_{pm} denote the weighted mean density and the heat capacity of the solid catalyst and the fluid, respectively, and D_{ea} and λ_{ea} denote the effective axial dispersion for axial mass and heat transport, respectively. Catalyst deactivation may be written as a decrease in the rate constant, which is often a function of temperature:

$$\frac{dk(T,t)}{dt} = f(k,T). \tag{5.11}$$

When calculating the instantaneous outlet concentrations due to fast variations in inlet concentrations, the temperature and the catalyst activity do not change, and only Equation (5.9) is needed if the temperature and the catalyst activity are known. When solving for bed temperature on the minutes scale, the catalyst activity can be assumed constant and the species can be assumed to be at steady state. Setting the activity constant means that the time derivative in Equation (5.11) is zero, and setting the right-hand side in the species balance to zero means that the species are at steady state and the accumulation term in the material balance can be omitted. Note the difference in the equations when the right-hand side is zero and when the left-hand side is zero. When the right-hand side is zero, the time derivative is zero, and nothing will change in time; when the left-hand side is zero, we assume negligible accumulation, and the dependent variable will immediately reach its steady-state value.

The time for reaching steady state for a process with stationary boundary conditions is estimated using the time constants. We can estimate the importance of the accumulation term in Equation (5.9) by comparing the accumulation term with the dominating terms in the equation, i.e. convection or reaction. For simplicity, we can choose convection. We want to estimate how long it will take to reach steady state. At that time, the accumulation of species should be negligible:

$$\frac{\partial C}{\partial t} \ll v \frac{\partial C}{\partial x}.$$

Approximate

$$\frac{\partial C}{\partial t} \approx \frac{\Delta C}{\Delta t} \approx -\frac{C}{\tau} \quad \text{and} \quad \frac{\partial C}{\partial x} \approx \frac{\Delta C}{\Delta x} \approx -\frac{C}{L}.$$

The preceding condition will result in

$$\frac{C}{\tau} \ll v \frac{C}{L} \quad \text{or} \quad \tau \gg \frac{L}{v}.$$

The same discussion for Equation (5.10) yields

$$\rho_m c_{pm} \frac{T}{\tau} \ll v \rho c_p \frac{T}{L} \quad \text{or} \quad \tau \gg \frac{\rho_m c_{pm} L}{\rho c_p v}.$$

Steady state for concentration is reached when $\tau \gg L/v$, and for temperature when $\tau \gg \rho_m c_{pm} L / \rho c_p v$. For gas-phase reactions, the heat capacity of the bed is much larger than the heat capacity for the gas, i.e. $\rho_m c_{pm} \gg \rho c_p$, and the time constant for heat is much larger than the time constant for species. The densities of liquids and solids are more equal, and the difference between the time constants for species and heat is
less.

An analysis of the time constant reveals that the first approximation, that change in concentration is much faster than change in temperature, may be a correct assumption for gas flow, but this is questionable for liquid flow due to the difference in heat capacity.

5.1.6   Linearizing

Linear equations are much easier to solve. Many problems that describe laminar flow or diffusion in simple geometry can be solved analytically if the mathematical model is linear. In many cases, the equations are basically linear, but small temperature- and concentration-dependent parameters, e.g. viscosity, density, and diffusivity, make the equations non-linear. A small non-linearity may have a minor influence on the accuracy and speed of a numerical solution, but it prevents an analytical solution. Assuming constant parameters over a volume and solving the equations analytically, and later iterating the solution with varying parameters, may be a very effective way of solving difficult equations. Even very simple linear approximations of the true model are useful for estimating order of magnitude.

Example 5.2 Linearizing a source term

The conversion of a second-order reaction in a tubular reactor with axial dispersion,

$$D_{ea}\frac{d^2C}{dx^2} - v\frac{dC}{dx} - kC^2 = 0 \tag{5.12}$$

can be overestimated by replacing the second-order reaction by a first-order reaction using the maximal concentration in the reactor,

$$D_{ea}\frac{d^2C}{dx^2} - v\frac{dC}{dx} - kC^{IN}C = 0, \tag{5.13}$$

where C^{IN} is the inlet concentration. A better approximation is obtained if C^{IN} is replaced by the average concentration, calculated from the solution of the first simple approximation with C^{IN}. Using the average concentration will underestimate the conversion in the first part of the reactor by giving a reaction rate at the inlet that is too low and consequently overestimating the reaction rate in the rear part.

More generally a simple form of linearization is a Taylor expansion around a given value a for the variable x:

$$f(x) = f(a) + \frac{f'(a)}{1!}(x - a) + \frac{f''(a)}{2!}(x - a)^2 + \frac{f'''(a)}{3!}(x - a)^3 + \cdots$$

$$= \sum_{n=0}^{\infty} \frac{f^{(n)}(a)}{n!}(x - a)^n, \tag{5.14}$$

and by evaluating the omitted terms it is possible to determine whether the linearization is an over- or an underestimation.

Example 5.3 Simplified pendulum motion

The equation for a pendulum is given by

$$\frac{d^2\omega}{dt^2} = -\frac{g}{L}\sin(\omega). \tag{5.15}$$

A linearization of the model using a Taylor expansion of $\sin(\omega)$ around $\omega = 0$ yields $\sin(\omega) = \omega - \omega^3/3! + \omega^5/5! + \cdots$. Keeping the linear term ω gives the linear differential equation

$$\frac{d^2\omega}{dt^2} = -\frac{g}{L}\omega, \tag{5.16}$$

which has the solution

$$\omega(t) = \omega_0 \cos\left(\sqrt{\frac{g}{L}}t\right), \qquad \omega \ll 1. \tag{5.17}$$

The omitted term $\omega^3/3!$ has the opposite sign to ω, and the forces on the right-hand side will be overestimated using Taylor linearization. The positive third term will hardly ever have an effect because the maximum angle is π due to the physics of the pendulum.

5.1.7   Limiting cases

Solving for limiting cases is always a good starting point. If we can estimate an upper and a lower bound for our calculations, we can easily see when our complete solution gives unreasonable results. Sometimes the upper and lower limits are both within the wanted precision, and no further calculations are necessary.

Example 5.4 Limiting cases for free body motion

An object is falling in air according to

$$m\frac{dv}{dt} + mg - C_D A \frac{\rho v^2}{2} = 0. \tag{5.18}$$

Two limiting cases can be found.

(i) At low velocity, the gravity is much larger than the drag resistance, resulting in

$$mg \gg C_D A \frac{\rho\, v^2}{2}. \tag{5.19}$$

Neglecting the term $C_D A(\rho\, v^2/2)$ in Equation (5.18) yields

$$\frac{dv}{dt} + g = 0 \tag{5.20}$$

and

$$v = v_{init} + gt. \tag{5.21}$$

(ii) After a long period of time, the object will reach a constant terminal velocity, resulting in the acceleration approaching zero:

$$\frac{dv}{dt} \to 0. \tag{5.22}$$

Neglecting the term dv/dt in Equation (5.18) yields

$$mg - C_D A \frac{\rho v^2}{2} = 0 \tag{5.23}$$

and

$$v_{term} = \sqrt{\frac{2mg}{C_D A \rho}}. \tag{5.24}$$

The first approximation gives a velocity that is too high after a long period of time, and the second gives a velocity that is too high after a short period of time. A combination of Equations (5.21) and (5.24), using the first one until the terminal velocity is reached and using the terminal velocity after that, will give a good

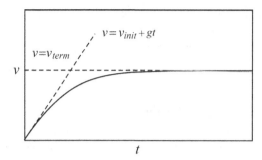

Figure 5.4. Velocity of a body falling in air.

approximation. However, the solution still overestimates the velocities and the distance the body has fallen, as shown in Figure 5.4.

5.1.8   Neglecting terms

The desired accuracy in prediction is seldom the maximal accuracy that can be obtained numerically. A good result is if a reactor volume can be estimated to within 10%. Often the calculated reactor volume is increased by 20% to 50% in order to compensate for uncertainties in rate constants and other parameters. In many cases, we can eliminate the terms in the equations that contribute only a few percent to the final result. If we can estimate whether omitting the terms will give an over- or underestimation of the final result, we can compensate for this uncertainty.

5.1.8.1   Order-of-magnitude estimation of derivatives

First we need some tools for analyzing differential equations. The following method is very simple and useful, but gives only an order-of-magnitude estimation of the terms in a differential equation. It is always necessary to verify that the assumptions were justified after the equations have been solved and more accurate estimations of the derivatives are possible.

In a general differential equation, it may be difficult to estimate the influence of different terms. One approach is to assume a solution and numerically approximate the derivatives. After solving the simplified equation by canceling unimportant terms, the initial assumptions can be confirmed.

A first-order derivative can be approximated by

$$\frac{dC}{dz} \approx \frac{\Delta C}{\Delta z},$$

and the second-order derivative can be approximated by

$$\frac{d^2 C}{dz^2} \approx \frac{\left.\frac{dC}{dz}\right|_{z+dz} - \left.\frac{dC}{dz}\right|_{z}}{dz} \approx \frac{\left.\frac{\Delta C}{\Delta z}\right|_{L} - \left.\frac{\Delta C}{\Delta z}\right|_{0}}{L}.$$

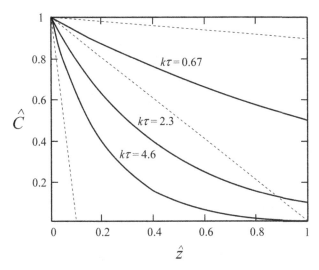

Figure 5.5. Approximate concentration profiles (solid lines) in the tubular reactor at different residence times or rate constants, simulated without the axial dispersion term. The dashed lines indicate slopes of 0.1, 1, and 10, respectively.

Example 5.5 Axial dispersion for a tubular reaction

Find a criterion for neglecting the axial dispersion of a first-order reaction in a tubular reactor. A first-order reaction in a tubular reactor without dispersion is described by

$$D_{ea} \underbrace{\frac{d^2 C}{d z^2}}_{\text{dispersion}} - \underbrace{U \frac{d C}{d z}}_{\text{convection}} - \underbrace{kC}_{\text{reaction}} = 0. \tag{5.25}$$

In the dimensionless form,

$$\frac{1}{Pe} \frac{d^2 \hat{C}}{d \hat{z}^2} - \frac{d \hat{C}}{d \hat{z}} - k\tau \hat{C} = 0, \tag{5.26}$$

with $\tau = L/U$, and the boundary conditions

$$\hat{C} = 1 \quad \text{at} \quad \hat{z} = 0$$

and

$$\frac{d\hat{C}}{d\hat{z}} = 0 \quad \text{at} \quad \hat{z} = 1.$$

First we need an order-of-magnitude estimation of the derivatives. A preliminary assumption is that the concentration profile has a shape as shown in Figure 5.5, i.e. the outlet concentration is much lower than the inlet concentration, but not zero. This assumption must be verified against the final model.

A coarse estimation of the derivatives is for the first-order derivative

$$\frac{dC}{dz} \approx \frac{\Delta C}{\Delta z} = \frac{0 - C^{IN}}{L - 0} = -\frac{C^{IN}}{L}, \tag{5.27}$$

in the dimensionless form

$$\frac{d\hat{C}}{d\hat{z}} \approx \frac{\Delta\hat{C}}{\Delta\hat{z}} = -\frac{1}{1} = -1; \tag{5.28}$$

and for the second-order derivative

$$\frac{d^2C}{dz^2} \approx \frac{\left.\frac{dC}{dz}\right|_{z+dz} - \left.\frac{dC}{dz}\right|_z}{dz} \approx \frac{\left.\frac{\Delta C}{\Delta z}\right|_L - \left.\frac{\Delta C}{\Delta z}\right|_0}{L} = \frac{0 - \left(-\frac{C^{IN}}{L}\right)}{L} = \frac{C^{IN}}{L^2}, \tag{5.29}$$

in the dimensionless form

$$\frac{d^2\hat{C}}{d\hat{z}^2} \approx \frac{\left.\frac{\Delta\hat{C}}{\Delta\hat{z}}\right|_1 - \left.\frac{\Delta\hat{C}}{\Delta\hat{z}}\right|_0}{1} = \frac{0 - (-1)}{1} = 1, \tag{5.30}$$

using the boundary condition $d\hat{C}/d\hat{z} = 0$ at $\hat{z} = 1$.

We can now make an order-of-magnitude estimation of the relative importance of the different terms in (5.26), repeated here for convenience:

$$\frac{1}{Pe}\frac{d^2\hat{C}}{d\hat{z}^2} - \frac{d\hat{C}}{d\hat{z}} - k\hat{C}\tau = 0 \tag{5.26}$$

$$\frac{1}{Pe} \qquad 1 \qquad k\tau$$

The axial dispersion term can be neglected when it is small in comparison with the largest terms in the model, i.e.

$$\frac{1}{Pe} \ll 1 \approx k\tau. \tag{5.31}$$

It follows from the transport equation (5.26) that $k\tau \approx 1$ if the axial dispersion can be neglected. If this criterion is fulfilled, Equation (5.26) can be approximated by

$$\frac{d\hat{C}}{d\hat{z}} = -k\hat{C}\tau. \tag{5.32}$$

This is a first-order ordinary differential equation. For this type of equation, we need only one boundary condition. In this case, the inlet condition $\hat{C} = 1$ at $\hat{z} = 0$ is the most important.

The solution of Equation (5.32) is

$$\hat{C} = e^{-k\tau\hat{z}}. \tag{5.33}$$

An analysis of the validity of the approximations is always necessary, and the first and second derivatives are calculated by differentiating Equation (5.33) to obtain

$$\frac{d\hat{C}}{d\hat{z}} = -k\tau e^{-k\tau\hat{z}} \tag{5.34}$$

and

$$\frac{d^2\hat{C}}{d\hat{z}^2} = k^2\tau^2 e^{-k\tau\hat{z}}. \tag{5.35}$$

Table 5.1. Conversion, first- and second-order derivatives at the inlet and outlet for different residence times

$k\tau$	$\hat{C}_{out}$	$-\dfrac{d\hat{C}}{d\hat{z}}\bigg|_{\hat{z}=0}$	$-\dfrac{d\hat{C}}{d\hat{z}}\bigg|_{\hat{z}=1}$	$\dfrac{d^2\hat{C}}{d\hat{z}^2}\bigg|_{\hat{z}=0}$	$\dfrac{d^2\hat{C}}{d\hat{z}^2}\bigg|_{\hat{z}=1}$
0.69	0.5	0.69	0.35	0.48	0.24
2.3	0.1	2.3	0.23	5.3	0.53
4.6	0.01	4.6	0.046	21.2	0.21

The results from the approximative solutions in Equations (5.33)–(5.35) are estimated for different parameters $k\tau$ in Table 5.1.

The example shows that, for a reactor with a reasonable conversion between 50% and 99% ($\hat{C}_{out}$ between 0.5 and 0.01), the approximations

$$-\frac{d\hat{C}}{d\hat{z}}\bigg|_{\hat{z}=0} \approx 1 \quad \text{and} \quad \frac{d^2\hat{C}}{d\hat{z}^2}\bigg|_{\hat{z}=0} \approx 1$$

are within an order of magnitude of the calculated derivatives, except for the second-order derivative at the inlet at high conversion. The criterion that we can neglect axial dispersion when $D_{ea}/vL \ll 1$ can, for high conversion, be too weak in some parts of the reactor. The criterion

$$\frac{D_{ea}}{vL}\frac{d^2y}{dx^2} \ll \left|\frac{dy}{dx}\right| \approx \frac{L}{v}ky,$$

where the derivatives are calculated from the simplified Equation (5.32), should always be valid. We also know that $(D_{ea}/vL)(d^2y/dx^2) > 0$, and consequently removing that term will result in an overestimation of the conversion.

5.1.9   Changing the boundary conditions

The boundary conditions always affect the solution to an equation, but in many cases the boundary conditions can be reformulated such that the equation can be solved more easily analytically or numerically. In many cases, the boundary conditions are not obvious and have been introduced only to obtain an analytical solution to an equation.

Example 5.6 Boundary condition for a tubular reactor with axial dispersion

This scenario has been discussed for more than 50 years. The condition

$$\frac{dC}{dz} = 0 \quad \text{at} \quad z = L$$

is under dispute because it introduces a feedback of information from the end of the reactor that does not occur in real reactors. A more realistic formulation for an irreversible reaction is perhaps

$$C \to 0 \quad \text{when} \quad z \to \infty.$$

For diffusion and reaction in a porous spherical catalyst, the condition

$$\frac{dC}{dr} = 0 \quad \text{at} \quad r = 0$$

is not necessary. A boundary condition that does not allow diffusive transport through the center of the catalyst may be sufficient, i.e.

$$r^2 \frac{dC}{dr} \to 0 \quad \text{when} \quad r \to 0,$$

and a sufficient condition is that the derivative dC/dr is limited when $r \to 0$. In addition, the condition $C = 0$ at a distance δ from the surface could even be appropriate for fast reactions.

5.2   Case study: Modeling flow, heat, and reaction in a tubular reactor

The methods listed in Sections 5.1.1–5.1.7 are illustrated by a tubular reactor with a first-order reaction and laminar flow. Models for species, heat, and momentum have been formulated and simplified. In addition to showing the methods, we discuss the assumptions in a traditional 1D lumped-parameter model for a tubular reactor with axial dispersion,

$$\frac{\partial C}{\partial t} = D_{ea} \frac{\partial^2 C}{\partial z^2} - v \frac{\partial C}{\partial z} - k_n(T, t) \cdot C^n, \tag{5.36}$$

$$\rho_m c_m \frac{\partial T}{\partial t} = \lambda_{ea} \frac{\partial^2 T}{\partial z^2} - v \rho c_p \frac{\partial T}{\partial z} - k_n(T, t)(-\Delta H) \cdot C^n. \tag{5.37}$$

5.2.1   General equation for a cylindrical reactor

The following equations describe transport of species, heat, and momentum in a cylindrical tubular reactor with laminar flow

Material balance for species A

$$\frac{\partial C_A}{\partial t} + v_z \frac{\partial C_A}{\partial z} + v_r \frac{\partial C_A}{\partial r} + \frac{v_\theta}{r} \frac{\partial C_A}{\partial \theta}$$
$$= D_A \left[\frac{1}{r} \frac{\partial}{\partial r} \left(r \frac{\partial C_A}{\partial r} \right) + \frac{\partial^2 C_A}{\partial z^2} + \frac{1}{r^2} \frac{\partial^2 C_A}{\partial \theta^2} \right] - R_A. \tag{5.38}$$

Heat balance

$$\rho c_p \left(\frac{\partial T}{\partial t} + v_z \frac{\partial T}{\partial z} + v_r \frac{\partial T}{\partial r} + \frac{v_\theta}{r} \frac{\partial T}{\partial \theta} \right)$$
$$= k \left[\frac{1}{r} \frac{\partial}{\partial r} \left(r \frac{\partial T}{\partial r} \right) + \frac{\partial^2 T}{\partial z^2} + \frac{1}{r^2} \frac{\partial^2 T}{\partial \theta^2} \right] + Q_A. \tag{5.39}$$

Balances for axial, z, radial, r, and tangential, θ, momentum

$$\rho \left(\frac{\partial v_z}{\partial t} + v_r \frac{\partial v_z}{\partial r} + v_z \frac{\partial v_z}{\partial z} + \frac{v_\theta}{r} \frac{\partial v_z}{\partial \theta} \right) = \mu \left[\frac{1}{r} \frac{\partial}{\partial r} \left(r \frac{\partial v_z}{\partial r} \right) + \frac{\partial^2 v_z}{\partial z^2} + \frac{1}{r^2} \frac{\partial^2 v_z}{\partial \theta^2} \right]$$
$$+ \rho G_z - \frac{\partial p}{\partial z}; \tag{5.40}$$

$$\rho \left(\frac{\partial v_r}{\partial t} + v_r \frac{\partial v_r}{\partial r} + v_z \frac{\partial v_r}{\partial z} + \frac{v_\theta}{r} \frac{\partial v_r}{\partial \theta} - \frac{v_\theta^2}{r} \right)$$
$$= \mu \left[\frac{\partial}{\partial r} \left(\frac{1}{r} \frac{\partial}{\partial r} (r v_r) \right) + \frac{\partial^2 v_r}{\partial z^2} + \frac{1}{r^2} \frac{\partial^2 v_z}{\partial \theta^2} - \frac{2}{r} \frac{\partial v_z}{\partial \theta} \right] + \rho G_r - \frac{\partial p}{\partial r}; \tag{5.41}$$

$$\rho \left(\frac{\partial v_\theta}{\partial t} + v_r \frac{\partial v_\theta}{\partial r} + v_z \frac{\partial v_\theta}{\partial z} + \frac{v_\theta}{r} \frac{\partial v_\theta}{\partial \theta} - \frac{v_r v_\theta}{r} \right)$$
$$= \mu \left[\frac{\partial}{\partial r} \left(\frac{1}{r} \frac{\partial}{\partial r} (r v_\theta) \right) + \frac{\partial^2 v_\theta}{\partial z^2} + \frac{1}{r^2} \frac{\partial^2 v_z}{\partial \theta^2} + \frac{2}{r} \frac{\partial v_z}{\partial \theta} \right] + \rho G_\theta - \frac{1}{r} \frac{\partial p}{\partial \theta}. \tag{5.42}$$

Equation of continuity

$$\frac{\partial \rho}{\partial t} + \frac{1}{r} \frac{\partial}{\partial r} (\rho r v_r) + \frac{1}{r} \frac{\partial}{\partial \theta} (\rho v_\theta) + \frac{\partial}{\partial z} (\rho v_z) = 0. \tag{5.43}$$

5.2.1.1 Boundary conditions

At the inlet, we have only axial velocity, and no radial or tangential variation in concentration or temperature:

$$v_z = v_{av}, \quad v_r = 0, \quad v_\theta = 0, \quad C = C^{IN}, \quad T = T^{IN}.$$

At the outlet, we have only axial velocity and axial convection of species and heat:

$$v_r = 0, \quad v_\theta = 0, \quad \frac{\partial v_z}{\partial z} = 0, \quad \frac{\partial C}{\partial z} = 0, \quad \frac{\partial T}{\partial z} = 0.$$

We have no slip at the wall and no penetration of species, but heat transfer through the wall will occur:

$$v_z = 0, \quad v_r = 0, \quad v_\theta = 0, \quad \frac{\partial C}{\partial r} = 0, \quad k \frac{\partial T}{\partial r} = u(T - T_w).$$

The boundary conditions are not obvious; they require detailed knowledge of the actual flow pattern, and, as discussed, the outlet conditions are questionable.

5.2.2 Reducing the number of independent variables

With no density differences in the fluid, there will be no source term for the radial and tangential velocities. If the inlet and boundary conditions are axisymmetric, there will be no variation in the θ direction. Removing the tangential variations will allow us to reduce the simulation from 3D to 2D. The symmetric boundary condition at the center of the tube can be formulated as

$$\frac{\partial v_r}{\partial r} = 0, \quad \frac{\partial v_z}{\partial r} = 0, \quad \frac{\partial C}{\partial r} = 0, \quad \frac{\partial T}{\partial r} = 0 \quad \text{at} \quad r = 0.$$

With no variation in tangential direction, the balance equation for tangential momentum, Equation (5.42), and the θ dependence in the remaining Equations (5.38)–(5.43) should be removed.

Radial velocity will occur due to the inlet velocity, which has not taken the final parabolic form. A general rule of thumb is that the variation in radial velocity that arises from an uneven inlet can be neglected in a pipe after an inlet distance of δ, estimated from the pipe diameter and the Reynolds number:

$$\frac{\delta}{d} = 0.05 \, \dot{R}e.$$

5.2.3   Steady state or transient?

How fast the concentration or temperature will change due to a step change at an inlet gives us an idea about the time resolution that is required for a transient solution, and how quickly the reactor will reach steady state. One way to retrieve the time constants is to compare the accumulation term with the dominating transport term.

Axial transport: the ratio of accumulation to convection is written and simplified as

$$\frac{\partial C_A/\partial t}{v_z(\partial C_A/\partial z)} \approx \frac{\Delta C_A/\Delta t}{v_z(\Delta C_A/\Delta z)} \approx \frac{C_{A,in}\tau}{v_z(C_{A,in}/L)}.$$

This gives the time constant for axial convection as

$$\tau = \frac{C_{A,in}}{v_z \, C_{A,in}/L} = \frac{L}{v_z}.$$

Radial diffusion

$$\frac{\partial C_A/\partial t}{D\dfrac{1}{r}\dfrac{\partial}{\partial r}\left(r\dfrac{\partial C_A}{\partial r}\right)} \approx \frac{\Delta C_A/\Delta t}{D\dfrac{1}{R}\left(R\dfrac{\Delta C}{\Delta r^2}\right)} \approx \frac{C_A/\tau}{DC_A/R^2}.$$

This gives the time constant for radial diffusion as

$$\tau = C_A/(DC_A/R^2) = \frac{R^2}{D}.$$

As expected, it will take one residence time L/v_z for a change in inlet concentration to affect the whole length of the reactor, and it will take R^2/D for the concentration to diffuse from the center of the reactor to the wall in accordance with the time constants developed in Section 5.1.5.

Similar time constants can be estimated for momentum and heat transfer also. Table 5.2 shows the time constants for convective transport in the axial direction, the radial transport of momentum, mass, and heat, assuming $L = 10$ m, $v = 0.1$ m s^{-1} and laminar flow. The fluid properties that are typical for a gas are: density 1 kg m^{-3}, diffusivity 10^{-5} m^2 s^{-1}, heat capacity 1000 J kg^{-1} K^{-1}. Typical liquid properties are: density 1000 kg m^{-3}, diffusivity 10^{-9} m^2 s^{-1}, heat capacity 1000 J kg^{-1} K^{-1}, heat conductivity 0.5 W m^{-2} K^{-1}. The main differences between gases and liquids are in the

Table 5.2. Time constants of a 0.1 m s^{-1} gas and liquid flow in a 10 m tubular reactor with two different tube diameters, R

The fluid properties are typical for gas and liquid

Conditions	Axial transport, L/v_z	Radial momentum, $R^2\rho/\mu$	Radial diffusion, R^2/D	Radial heat conduction, $R^2\rho c_p/k$
Air (25 °C, 1 bar, $R = 0.1$ m)	100	500	1000	400
Water (25 °C, $R = 0.005$ m)	100	100	25 000	200

diffusivity and heat conduction. The very low diffusivity in liquids is, to some extent, compensated for in the table by a much smaller tube radius.

With a step response at the inlet and no external time-dependent influence, the reactor will reach steady state after the largest of the time constants. The much larger time constants for the radial diffusion (R^2/D) than for the residence time (L/v), indicate that the concentration will vary radially within the whole reactor if the inflow does not have equal concentration over the cross section.

In this case study we assume a constant even flow and concentration at the inlet and only consider steady state. However, the concentration will have a radial variation downstream due to the reaction and the radial difference in residence time. Assuming an even inflow and a minor effect of gravity in the radial and tangential directions, the convective radial transport terms can be removed. Axial viscous transport, diffusion, and conduction can also be neglected because the convective axial transport is much larger. After a sufficiently long period of time with no change in flow, the transient terms become very small, and we obtain a model with steady axial convection, radial diffusion, and conduction with a reaction source term.

Material balance

$$v_z \frac{\partial C_A}{\partial z} = D \frac{1}{r} \frac{\partial}{\partial r} \left(r \frac{\partial C_A}{\partial r} \right) - R_A. \tag{5.44}$$

Heat balance

$$\rho c_p \left(v_z \frac{\partial T}{\partial z} \right) = k \frac{1}{r} \frac{\partial}{\partial r} \left(r \frac{\partial T}{\partial r} \right) + Q_A. \tag{5.45}$$

Momentum balance

$$\rho \left(v_z \frac{\partial v_z}{\partial z} \right) = \mu \frac{1}{r} \frac{\partial}{\partial r} \left(r \frac{\partial v_z}{\partial r} \right) + \rho G_z - \frac{\partial p}{\partial z}. \tag{5.46}$$

Equation of continuity

$$\frac{\partial}{\partial z} (\rho v_z) = 0. \tag{5.47}$$

5.2.4   Decoupling equations

If, in the case study, μ and ρ are assumed to be independent of C_A and T in the actual concentration and temperature interval, the solution of the equations is drastically simplified. The equation of continuity gives $\partial v_z / \partial z = 0$, and using this in Equation (5.46) yields

$$\mu \frac{1}{r} \frac{\partial}{\partial r} \left(r \frac{\partial v_z}{\partial r} \right) = \frac{\partial p}{\partial z} - \rho G_z = \text{constant.} \tag{5.48}$$

Integrating Equation (5.48) and using $\partial v_z / \partial r = 0$ when $r = 0$ and $v_z = 0$ when $r = R$ yields

$$v_z(r) = \left(\rho G_z - \frac{\partial p}{\partial z} \right) \frac{R^2}{4\mu} \left(1 - \frac{r^2}{R^2} \right) = 2 v_{av} \left(1 - \frac{r^2}{R^2} \right), \tag{5.49}$$

where v_{av} denotes the average velocity in the pipe. As expected, we obtain the parabolic velocity profile for laminar flow in a circular pipe.

5.2.5   Simplified geometry

We have now reduced the equations to two coupled partial differential equations (using Equation (5.49) in Equations (5.44) and (5.45)):

$$2 v_{av} \left(1 - \frac{r^2}{R^2} \right) \frac{\partial C_A}{\partial z} = D \frac{1}{r} \frac{\partial}{\partial r} \left(r \frac{\partial C_A}{\partial r} \right) - R_A; \tag{5.50}$$

$$\rho c_p 2 v_{av} \left(1 - \frac{r^2}{R^2} \right) \frac{\partial T}{\partial z} = k \frac{1}{r} \frac{\partial}{\partial r} \left(r \frac{\partial T}{\partial r} \right) + Q_A. \tag{5.51}$$

The dimensionless space coordinates $\hat{r} = r/R$ and $\hat{z} = z/L$ yield

$$2 \left(1 - \hat{r}^2 \right) \frac{\partial C_A}{\partial \hat{z}} = \frac{L D}{R^2 v_{av}} \frac{1}{\hat{r}} \frac{\partial}{\partial \hat{r}} \left(\hat{r} \frac{\partial C_A}{\partial \hat{r}} \right) - \frac{L}{v_{av}} \cdot R_A; \tag{5.52}$$

$$2 \left(1 - \hat{r}^2 \right) \frac{\partial T}{\partial \hat{z}} = \frac{L \cdot k}{R^2 v_{av} \rho c_p} \frac{1}{\hat{r}} \frac{\partial}{\partial \hat{r}} \left(\hat{r} \frac{\partial T}{\partial \hat{r}} \right) + \frac{L}{v_{av} \rho c_p} \cdot Q_A. \tag{5.53}$$

In order to obtain ordinary differential equations, we must eliminate one space coordinate. There are several possibilities for doing this: we can calculate an average by integrating over the radial or axial coordinates; we can find simple polynomials that can approximate the radial concentration and temperature dependencies; or we can assume that the radial coupling is negligible.

The simplest solution is to use the radial average, as we assume that the axial variations are much larger than the radial variations.

First, we integrate radially (this mean to integrate over a thin circular slice $2\pi r \, dr$, which in the dimensionless form is $2\pi \hat{r} \, d\hat{r}$). The integration of the radial transport term is straightforward because

$$\int_0^1 \frac{d}{d\hat{r}}\left(\hat{r}\frac{dy}{d\hat{r}}\right)d\hat{r} = \hat{r}\frac{dy}{d\hat{r}}\bigg|_{\hat{r}=1} - \hat{r}\frac{dy}{d\hat{r}}\bigg|_{\hat{r}=0} = \hat{r}\frac{dy}{d\hat{r}}\bigg|_{\hat{r}=1}$$

and

$$\int_0^1 2\left(1-\hat{r}^2\right)\frac{\partial C_A}{\partial \hat{z}} 2\pi\hat{r}\,d\hat{r} = 2\pi\frac{L\,D}{R^2\,v_{av}}\frac{\partial C_A}{\partial \hat{r}}\bigg|_{\hat{r}=1} - \int_0^1 \frac{L}{v_{av}} R_A 2\pi\hat{r}\,d\hat{r}, \quad (5.54)$$

$$\int_0^1 2\left(1-\hat{r}^2\right)\frac{\partial T}{\partial \hat{z}} 2\pi\hat{r}\,d\hat{r} = 2\pi\frac{L\,k}{R^2\,v_{av}\,\rho\,c_p}\frac{\partial T}{\partial \hat{r}}\bigg|_{\hat{r}=1}$$

$$+ \int_0^1 \frac{L}{v_{av}\,\rho\,c_p} Q_A 2\pi\hat{r}\,d\hat{r}. \quad (5.55)$$

The wall is not permeable for mass transport, i.e.

$$\frac{\partial C_A}{\partial \hat{r}}\bigg|_{\hat{r}=1} = 0. \quad (5.56)$$

Heat is conducted through the wall, so

$$k\frac{\partial T}{\partial r}\bigg|_{r=R} = \frac{k}{R}\frac{\partial T}{\partial \hat{r}}\bigg|_{\hat{r}=1} = u(T_w - T_{x=1}). \quad (5.57)$$

We define the mixed-cup average $\overline{C}_A$ using

$$\overline{C}_A \cdot v_{av} = \frac{\int_0^1 2v_{av}(1-\hat{r}^2)\,C_A\,(\hat{z},\hat{r})\,2\pi\hat{r}\,d\hat{r}}{\int_0^1 2\pi\hat{r}\,d\hat{r}}$$

$$= \int_0^1 2v_{av}\left(1-\hat{r}^2\right)\,C_A\,(\hat{z},\hat{r})\,2\hat{r}\,d\hat{r} \quad (5.58)$$

and the mixed-cup average $\overline{T}$ using

$$\overline{T}\,v_{av} = \int_0^1 2\,v_{av}\left(1-\hat{r}^2\right)\,T\,(\hat{z},\hat{r})\,2\hat{r}\,d\hat{r}. \quad (5.59)$$

Observe that this averaging over the radius also eliminates the axial dispersion that is caused by the parabolic velocity profile.

Also note that the difference between the mixed-cup average and the average $\langle C \rangle$

$$\langle C_A \rangle = \frac{\int_0^1 C_A(\hat{z}, \hat{r}) 2\pi \hat{r} \, d\hat{r}}{\int_0^1 2\pi \hat{r} \, d\hat{r}} = \int_0^1 C_A(\hat{z}, \hat{r}) 2\hat{r} \, d\hat{r} \tag{5.60}$$

over the cylinder, is due to the radial variation in velocity. If we take a sample from the outlet to measure concentration, we will obtain more sample from the center of the pipe due to the higher velocity in the center. If we measure concentration spectrometrically, we will have equal contribution from all parts, irrespective of velocity. Usually it is the mixed-cup average we need as it is related to the total amount converted in the reactor.

The final integrals on the right-hand side of Equations (5.54) and (5.55) are approximated by

$$2 \int_0^1 \frac{L}{v_{av}} R_A (C_A, T) \hat{r} \, d\hat{r} \approx \frac{L}{v_{av}} R_A (\overline{C}_A, \overline{T}), \tag{5.61}$$

$$2 \int_0^1 \frac{L}{v_{av} \rho \, c_p} Q_A \hat{r} \, d\hat{r} \approx \frac{L}{v_{av} \, \rho \, c_p} Q_A (\overline{C}_A, \overline{T}), \tag{5.62}$$

i.e. the average of the reaction rate is approximated by the reaction rate at averaged concentration and temperature.

We now have two coupled ordinary differential equations:

$$\frac{d\overline{C}}{d\hat{z}} = -\frac{L}{v_{av}} R_A (\overline{C}_A, \overline{T}), \tag{5.63}$$

$$\frac{d\overline{T}}{d\hat{z}} = 2 \frac{L\bar{u}}{R \, v_{av} \rho \, c_p} (T_w - \overline{T}) + \frac{L}{v_{av} \, \rho \, c_p} Q_A (\overline{C}_A, \overline{T}). \tag{5.64}$$

However, comparing the heat flux through the wall, we have

$$u(T_w - T_{x=1}) = \bar{u}(T_w - \overline{T}), \tag{5.65}$$

but, because the mixed-cup temperature differs from the wall temperature, i.e.

$$\overline{T} \neq T_{\hat{r}=1},$$

we must adjust $\bar{u}$; one common method is to add a part of the heat transfer resistance in the reactor as follows:

$$\frac{1}{\bar{u}} = \frac{1}{u} + \alpha \frac{R}{k}. \tag{5.66}$$

By making this adjustment we assume that we have the temperature $\overline{T}$, a distance αR from the wall, and that heat is conducted through the fluid and the wall.

Because the equations are non-linear, there is still no analytical solution to the equations. The equations describe a simple initial-value problem, and numerical solutions are straightforward. An analytical solution is possible only if we assume that, by efficient cooling or through a reaction of low heat of reaction, the temperature will be constant axially and the reaction rate will be a function of concentration only, i.e.

$$\int_{C_A^{IN}}^{C_A^{OUT}} \frac{d\overline{C}}{\dfrac{L}{v_{av}} R_A(\overline{C}_A)} = -\int_0^1 d\hat{z}. \tag{5.67}$$

5.2.6 Limiting cases

The partial differential equation, Equation (5.52), can be approximated by using different assumptions. For an isothermal case, it can be written, for a first-order reaction, as follows:

$$2(1-\hat{r}^2)\frac{\partial C}{\partial \hat{z}} = \frac{DL}{R^2 v_{av}} \frac{1}{\hat{r}} \frac{\partial}{\partial \hat{r}} \hat{r} \frac{\partial C}{\partial \hat{r}} - \frac{Lk_1}{v_{av}} C. \tag{5.68}$$

(1) Assuming no radial variations. When the residence time is much larger than the time for radial diffusion, i.e.

$$\frac{L}{v_{av}} \gg \frac{R^2}{D}, \tag{5.69}$$

the ratio between residence time and the radial diffusion time constant appears in the term for radial diffusion, and

$$\frac{DL}{R^2 v_{av}} \rightarrow \infty$$

$$\Rightarrow \frac{\partial C}{\partial \hat{r}} \rightarrow 0$$

$$\Rightarrow C(\hat{r}, \hat{z}) = \overline{C}(\hat{z});$$

i.e. the radial concentration is assumed to be constant, equal to the mixed-cup average, as shown by eliminating radial variation. Integrating the average flow,

$$\int_0^1 2(1-\hat{r}^2)\frac{\partial C}{\partial \hat{z}} 2\hat{r} \, d\hat{r} = \frac{d\overline{C}}{d\hat{z}}, \tag{5.70}$$

will simplify the model to

$$\frac{d\overline{C}}{d\hat{z}} = -\frac{Lk_1}{v_{av}}\overline{C}, \tag{5.71}$$

$$\overline{C} = \exp[-(Lk/v_{av})\hat{z}]. \tag{5.72}$$

(2) Assuming no radial transport, i.e. a very short residence time in comparison to the time for radial diffusion, i.e.

$$\frac{R^2}{D} \gg \frac{L}{v_{av}}, $$

will remove the radial diffusion term and Equation (5.68) will be simplified to

$$2(1-\hat{r}^2)\frac{\partial C}{\partial \hat{z}} = -\frac{Lk_1}{v_{av}} \cdot C. \tag{5.73}$$

Rearranging this equation yields

$$\frac{\partial C}{C} = -\frac{L\,k_1}{2\cdot v_{av}(1-\hat{r}^2)}\,\partial\hat{z}$$ (5.74)

and the solution

$$C(\hat{z},\hat{r}) = C_{in}\exp\left[-\frac{L\cdot k_1}{2\cdot v_{av}(1-\hat{r}^2)}\cdot\hat{z}\right].$$ (5.75)

The concentration in the outlet will be a function of both radius and reactor length. A mixed-cup average is obtained by integrating the concentration according to Equation (5.58).

5.2.7 Conclusions

The complex partial differential equations have been simplified all the way, until an analytical solution is possible. The simplifications can be halted at any point depending on the desired accuracy and available numerical tools.

The assumptions in a lumped-parameter model are not always transparent. For example, in the 1D model for a tubular reactor with axial dispersion (Equations (5.36) and (5.37), repeated here for convenience)

$$\frac{\partial C}{\partial t} = D_{ea}\frac{\partial^2 C}{\partial z^2} - v\frac{\partial C}{\partial z} - k_n(T,t)\cdot C^n,$$ (5.36)

$$\rho c_p\frac{\partial T}{\partial t} = \lambda_{ea}\frac{\partial^2 T}{\partial z^2} - v\rho c_p\frac{\partial T}{\partial z} - k_n(T,t)(-\Delta H)\cdot C^n,$$ (5.37)

where the axial dispersion term D_{ea} can be estimated from

$$D_{ea} = D + \frac{R^2 v^2}{192 D},$$

and λ_{ea} by replacing D by thermal diffusion α in the equation.

The main assumptions in the lumped-parameter model are

(1) the average reaction rate is the rate at average temperature and concentration;
(2) the axial dispersion term describes the complex combination of transport by velocity gradients and diffusion;
(3) the velocity is the traditional average, and the concentration and temperature are the mixed-cup averages.

5.3 Error estimations

The simulations must be validated and verified before they can be trusted. At the beginning of this chapter, we stated that poor simulation results may be due to many different reasons: errors in formulating the problem, inaccurate models, inaccurate data on physical properties, numerical errors due to the choice of numerical methods, and lack of convergence in the simulations.

5.3.1 Sensitivity analysis

Validating the model by comparing it with the experimental data of a similar system is the most direct way to test all the aforementioned errors. However, it is not always obvious what a difference between simulations and experiments indicates, and it is important to try to evaluate each possible error separately. Errors in formulating the problem may be recognized from simulations of limiting cases, e.g. very high or low kinetic rate constants or velocities. Simulation of an isothermal case will remove errors in the heat balance, and simulation using a preset velocity profile will remove errors in the momentum balance, etc. Inaccurate physical data can be evaluated by making a small change in data values and estimating how sensitive the results are to the absolute value of the variables. This type of analysis will tell us which variables are critical and must be known very accurately.

In addition to changing parameters to determine the influence of different phenomena, the influence of the different terms in a model can be estimated by putting a weighting factor in front of each term in the model, e.g. the accumulation term or the axial dispersion term in a chemical reactor model, and comparing the simulations. In this way, we can easily estimate which factors dominate the model, and perhaps remove the parts that have negligible influence. We can also judge whether the result supports our expectations or whether the model must be reformulated.

Numerical errors and convergence can be tested by changing the initial guess of the solution and the numerical accuracy in the numerical subroutines. In an optimization or least squares minimization using a gradient search, it is important to have high accuracy in order to calculate a correct gradient. The step length in a gradient evaluation must be chosen sufficiently large in comparison with the solution accuracy. A gradient-free search, e.g. a random Monte Carlo method, may be more efficient.

5.3.2 Over- and underestimations

A sensitivity analysis will reveal how variables affect final results. We can find out, e.g., whether our estimated chemical reactor volume will increase or decrease when a certain parameter is increased. If that parameter is uncertain, we can simulate for an upper and a lower value of that parameter and obtain an over- and an underestimation of, e.g., the reactor volume. The same method can be used for other uncertainties in modeling, e.g. to compare results assuming laminar or turbulent flow, and whether to include radiation or not.

5.4 Questions

(1) What properties are essential in an equation that may be solved independently of the remaining equations in a system of equations?

(2) Why are the net mass and heat transfer through a symmetry boundary condition zero?

(3) What are the maximum number of symmetry planes in a cube, cylinder, and sphere?

(4) In Equation (5.9), what is the difference if the left-hand side or the right-hand side is set to zero?

(5) How many symmetry planes exist for a cube exposed to gravity? Analyze the cases with gravity along the sides and along the diagonal of the cube.

5.5   Practice problems

5.1 Estimate how far into a liquid a compound will diffuse or heat will conduct in 1 min. (Diffusivity $= 10^{-9}$ m^2 s^{-1}, $\rho = 1000$ kg m^{-3}, $c_p = 4000$ J kg^{-1}, $k = 0.2$ J m^{-1} K^{-1}.) Use $\hat{y} = 0.16$.

5.2 About 150 years ago, Lord Kelvin estimated the age of the earth to be about 100 million years. He assumed that the earth originally had a temperature of 6000 °C and that it now has an average temperature of 3000 °C. He solved the heat equation for the different zones,

$$\rho c_p \frac{\partial T}{\partial t} = k \left(\frac{\partial^2 T}{\partial x_1^2} + \frac{\partial^2 T}{\partial x_2^2} + \frac{\partial^2 T}{\partial x_3^2} \right),$$

with the boundary conditions

$$kn \left(\frac{dT}{dx_1} + \frac{dT}{dx_2} + \frac{dT}{dx_3} \right) = \sigma T^4,$$

where n is the normal to the surface and

$$\frac{dT}{dx_i} = 0$$

in the middle at $x_i = 0$.

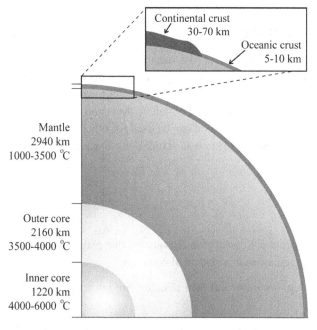

Figure P5.1

Discuss how the following methods can be used to simplify the calculations:
- decoupling of equations;
- reducing the number of independent variables;
- separation of equations into steady state and transient;
- simplifying geometry;
- linearizing;
- neglecting terms;
- solving for limiting cases.

Make your own estimation using the data in Figure P5.1. The radiation inflow from the sun and the radiation outflow are today the dominant terms; the solar constant is 1400 W m^{-2}. The net flux from the inner part of the earth is today on the order of $5 \cdot 10^{-2}$ W m^{-2}. Assume a heat capactity of $5 \cdot 10^6$ J m^{-3} K^{-1}.

5.3 Most of the harmful emissions from cars appear during cold start because the catalyst has to reach 250 °C before it becomes active. It takes about 1 min under normal driving conditions. An option to reduce the time is to put a smaller catalyst with less thermal mass close to the engine. Using late ignition in the engine makes it possible to raise the exhaust temperature very quickly. Calculate an over- and underestimation of how long it will take to heat the catalytic converter to 250 °C when the exhaust gas temperature increases linearly from 20 to 700 °C for 10 s and is then kept constant.

The catalyst has a volume of 0.5 l and a mass of 100 g. The channel surface (i.e. the area of contact with the exhaust gases) is 2000 m^2 m^{-3}. The mass flow of the exhaust gas is 0.01 kg s^{-1}. Assume that the catalyst heat capacity is 1000 J kg^{-1} K^{-1}, and that of the exhaust gas is 1000 J kg^{-1} K^{-1}. The Nusselt number for heat transfer in the channels is 4, and the gas heat conduction is 0.04 W K^{-1} m^{-1}.

5.4 Over 80 times more CO_2 is dissolved in the sea than in the atmosphere, and the gas will not contribute to global warming if it dissolves in the oceans. The primary mechanism for the transport of CO_2 from the atmosphere to the ocean is through its dissolution in raindrops. Calculate approximately (over- and underestimates) how much it needs to rain in Sweden (486 000 km^2) in m m^{-2} if the production of CO_2 (55 million tons yr^{-1}) is balanced with the amount that dissolves in the rain. Assume that the raindrops are of about 6 mm diameter and that they fall with a velocity of 10 m s^{-1} from 1000 m.

The transport of the CO_2 into a raindrop can be approximated by diffusion:

$$\frac{\partial C}{\partial t} = D \left(\frac{\partial^2 C}{\partial r^2} + \frac{2}{r} \frac{\partial C}{\partial r} \right).$$

The atmospheric CO_2 concentration is 0.0325%, and its solubility in water is 40 mol m^{-3} bar^{-1}. The diffusivity of CO_2 in water is 10^{-9} m^2 s^{-1}. The model neglects the deformation of the droplet, the convection inside the droplet, and the mass transfer in the air outside the droplet. Does this lead to an over- or underestimation of CO_2 transport into the drop?

Hint: Estimate how far into the drop the CO_2 can diffuse in the available time.

5.5 The surface temperature in the desert can vary between $+80\ °C$ during daytime and $-10\ °C$ during the night. Estimate how far into the sand the temperature fluctuations can reach. Define the position where no fluctuation is observed as the point where less than 1% of the surface variations are noticed. The density of the sand is 2000 kg m^{-3}, the heat conductivity is 0.8 W m^{-1} K^{-1}, and the heat capacity is 1000 J kg^{-1} K^{-1}.

5.6 Monolithic catalytic converters with a large number of parallel channels are very common in emission control for cars. The flow is laminar in order to minimize pressure drop with parabolic velocity profile. The maximum velocity in the center is twice the average velocity. Calculate the required catalyst volume assuming the rate-determining step is that the pollutants should have time to diffuse from the center of the channel to the catalytic wall. The channels are 1 mm wide and the channel volume is 80% of the catalytic converter volume. The volumetric flow of exhaust gas for a 2 l engine running at 3000 rpm can be approximated as (engine speed) × (engine volume)/2.

6 Numerical methods

Differential equations play a dominant role in mathematical modeling. In practical engineering applications, only a very limited number of them can be solved analytically. The purpose of this chapter is to give an introduction to the numerical methods needed to solve differential equations, and to explain how solution accuracy can be controlled and how stability can be ensured by selecting the appropriate methods. The mathematical framework needed to solve both ordinary and partial differential equations is presented. A guideline for selecting numerical methods is presented at the end of the chapter.

6.1 Ordinary differential equations

A characteristic of a differential equation is that it involves an unknown function and one or more of the function's derivatives. If the unknown function depends on only one independent variable, it is classified as an ordinary differential equation (ODE). The order of the differential equation is simply the order of the highest derivative that appears in the equations. Consequently, a first-order ODE contains only first derivatives, whilst a second-order ODE may contain both second and first derivatives. The ODEs can also be classified as linear or non-linear. Linear ODEs are the ones in which all dependent variables and their derivatives appear in a linear form. This implies that they cannot be multiplied or divided by each other, and they must be raised to the power of 1. An ODE has an infinite number of solutions, but with the appropriate conditions that describe systems, i.e. the initial value or the boundary value, the solutions can be determined uniquely.

6.1.1 ODE classification

ODEs can be classified as initial-value problems (IVPs), or boundary-value problems (BVPs). The classification into IVP or BVP depends on the location of the extra conditions specified. For IVPs they are given at the same value of the independent variable (the lower boundary, and consequently the initial value), whereas for BVPs they are given at different values of the independent variable. Consider Equation (6.1), a second-order

ODE on an interval $x \in [a, b]$, with two different sets of conditions, which cause the equation to be classsified as either IVP or BVP:

$$\frac{d^2 y}{dx^2} = f\left(x, y, \frac{dy}{dx}\right). \tag{6.1}$$

Initial-value problem

$$y(a) = \alpha,$$
$$\left.\frac{dy}{dx}\right|_{x=a} = \beta.$$

Boundary-value problem

$$y(a) = \alpha,$$
$$y(b) = \gamma.$$

The numerical methods used to solve these two classes of problems, i.e. IVPs and BVPs, are different, and the two separate conditions for the BVP generally make it more difficult to solve. Methods for solving both of these types of problems will be introduced in this chapter.

6.1.2   Solving initial-value problems

In order to give an introduction to the most important concepts used in different methods, such as accuracy, stability, and efficiency, we will start with the explicit Euler method, which is the simplest numerical method to use when solving ODEs. At this point, we should stress that the accuracy of this method is low, and that it is only conditionally stable. For this reason, it is not used in practice to solve any engineering problems. However, it serves well to illustrate a method for solving ODEs numerically, and for introducing the concepts of accuracy and stability. Later in this chapter more advanced numerical methods will be presented; methods that have higher accuracy and better stability properties, and that are implemented in modern software products.

6.1.2.1   The explicit Euler method

Consider the solution of an initial-value problem

$$\frac{dy}{dx} = f(x, y), \tag{6.2}$$

with the initial condition $y(x_0) = y_0$. The approach used in the explicit Euler method is based on the discretization of the interval $x \in [a, b]$ into M equal subintervals, which gives the step size $h = (b - a)/M$. The explicit Euler method approximates the solution as

$$y_{n+1} = y_n + hf(x_n, y_n). \tag{6.3}$$

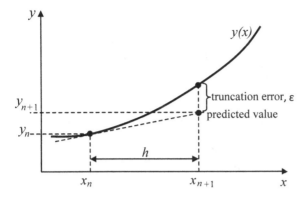

Figure 6.1. Principle of the Euler method.

The error made in the Euler method in a single step, ε, can be estimated by using the Taylor expansion

$$y(x_0 + h) = y(x_0) + hy'(x_0) + \frac{1}{2}h^2 y''(x_0) + O(h^3). \tag{6.4}$$

By subtracting Equation (6.3) from Equation (6.4) we obtain

$$\varepsilon = \frac{1}{2}h^2 y''(x_0) + O(h^3), \tag{6.5}$$

which shows that the local truncation error is approximately proportional to h^2. It can also be shown that the global error over the interval $x \in [a, b]$ is $O(h)$. (This will be dicussed in detail later in the chapter.) As shown in Figure 6.1, the derivative changes between x_n and x_{n+1}, but the Euler method simply relies on the derivative at x_n.

Example 6.1 Error versus step size
A simple problem illustrates how the error is related to step size. Consider the following problem:

$$\begin{cases} \dfrac{dy}{dt} = f(t, y) = y + t, \\ y(0) = 1, \\ t \in [0, 1]. \end{cases}$$

The numerical solution using the explicit first-order Euler method, and the step size $h = 0.2$, is shown in Figure 6.2.

By using the globally first-order accurate Euler method, $O(h)$, the error is halved becomes when the step size is halved. If the step size is further reduced, the global error continues to decrease, but very slowly, as shown in Figure 6.3. A major drawback of the Euler method is that it has low accuracy; a very small step size is required to control and keep error low. The errors, shown in Figure 6.3, have been calculated as the

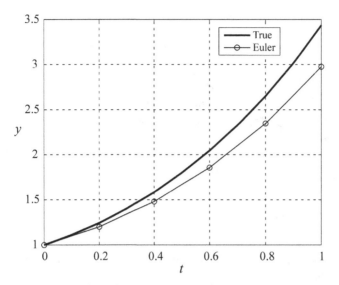

Figure 6.2. Solution using the Euler method, step size $h = 0.2$.

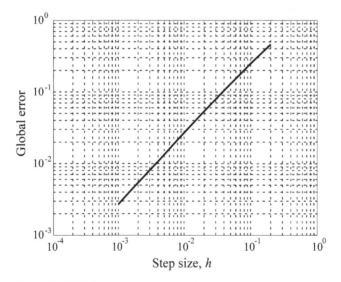

Figure 6.3. Global error vs. step size.

difference between the numerical solution and the true value. In this case, the error can be easily calculated as the analytical solution is given by $y(t) = 2\exp(t) - t - 1$.

6.1.2.2   The midpoint method

The midpoint method is an improvement on the Euler method in that it uses information about the function at a point other than the initial point of the interval. As seen in Figure 6.4, the values of x and y at the midpoint are used to calculate the step across the whole interval, h.

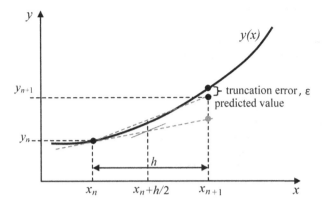

Figure 6.4. Principle of the midpoint method.

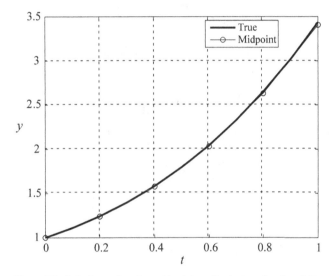

Figure 6.5. Solution using the midpoint method, step size $h = 0.2$.

The midpoint method requires two function evaluations in the step,

$$k_1 = hf(x_n, y_n), \tag{6.6}$$

$$k_2 = hf(x_n + h/2, y_n + k_1/2), \tag{6.7}$$

and gives the approximate solution as

$$y_{n+1} = y_n + k_2 + O(h^3). \tag{6.8}$$

Consequently, the midpoint method increases accuracy by one order. Obviously the increased order of accuracy leads to a better approximation than the Euler solution, as shown in Figure 6.5.

A comparison between the errors in the Euler and midpoint methods is summarized in Table 6.1 and shown in Figure 6.6.

Table 6.1. Comparison between errors in the Euler and midpoint methods

Steps, n	Error, e_i (Euler)	Error, e_i (midpoint)
5	0.4599	0.0311
10	0.2491	0.0084
20	0.1300	0.0022
50	0.0534	0.000357
100	0.0269	8.99e − 5
1000	0.0027	9.05e − 7

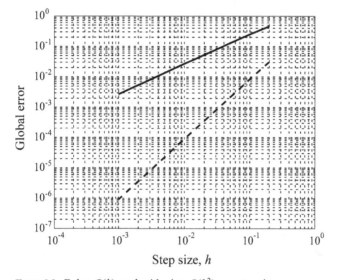

Figure 6.6. Euler, $O(h)$, and midpoint, $O(h^2)$, vs. step size.

6.1.2.3   Runge–Kutta (fourth order)

The fourth-order Runge–Kutta method is sometimes used because the method is accurate, stable, and easy to program. However, there are much better alternatives, as we will soon discover. This method uses four function evaluations per step,

$$k_1 = hf(x_n, y_n), \tag{6.9}$$

$$k_2 = hf(x_n + h/2, y_n + k_1/2), \tag{6.10}$$

$$k_3 = hf(x_n + h/2, y_n + k_2/2), \tag{6.11}$$

$$k_4 = hf(x_n + h, y_n + k_3), \tag{6.12}$$

and gives the approximate solution as

$$y_{n+1} = y_n + k_1/6 + k_2/3 + k_3/3 + k_4/6 + O(h^5). \tag{6.13}$$

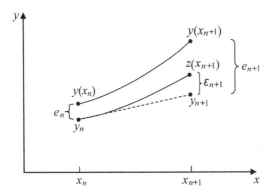

Figure 6.7. Local and global errors.

These methods (Euler, midpoint, and Runge–Kutta) are examples of explicit algorithms. For the same step size, the higher-order Runge–Kutta method gives better precision but at the cost of more calculations in each step.

6.1.3 Numerical accuracy

There are always errors present when we solve differential equations numerically. These errors can be divided into two different parts:

- a *local* truncation error, which is due to the numerical method in each step;
- a *global* error, which is due to previous local errors and amplification.

To be clear, let us first define the correct solution to the initial-value problem as $y(x)$. The approximative solution using a single-step numerical method, e.g. the Euler method, with an increment function $f(x_n, y_n)$ and step size h is $y_{n+1} = y_n + hf(x_n, y_n)$, and it is approximated at a number of discrete points $\{(x_n, y_n)\}_{n=0}^{M}$.

The global truncation error, e_n, is defined as the difference between the approximation, y_n, and the correct solution of the IVP, $y(x_n)$:

$$e_n = y_n - y(x_n). \tag{6.14}$$

The local error, ε_{n+1}, is defined as the error committed in each step from x_n to x_{n+1}. Let us introduce $z(x)$, which is the correct solution of the IVP over that single step. The local truncation error is then given by

$$\varepsilon_{n+1} = y_{n+1} - z(x_{n+1}). \tag{6.15}$$

Whereas the local error is due to a single step, the global error, e_{n+1}, is the sum of the local error, ε_{n+1}, and the amplified error from previous steps. As a consequence, it is not equal to the sum of the local errors from previous steps. The local and global errors are illustrated in Figure 6.7. Note that the upper curve, $y(x)$, is the correct (true) solution to the IVP and that the lower curve, $z(x)$, is the correct solution of the IVP on $x_n < x < x_{n+1}$, given $y(x_n) = y_n$.

Table 6.2. Summary of accuracy of the three single-step methods

Method	Local error	Global error
Euler	$O(h^2)$	$O(h)$
Midpoint (second-order Runge–Kutta)	$O(h^3)$	$O(h^2)$
Fourth-order Runge–Kutta	$O(h^5)$	$O(h^4)$

Generally, if the local discretization error is $O(h^n)$, the global error is $O(h^{n-1})$. Consequently, the Euler method is locally second-order accurate, $O(h^2)$, and globally (over the interval $x \in [a, b]$) it is first-order accurate, $O(h)$.

The accuracy of the three single-step methods introduced so far in this chapter is presented in Table 6.2. For practical applications, higher-order methods are needed to reduce the computational effort. Therefore, the advantage of using a higher-order method does not come from getting an extremely accurate solution; in most cases, it is satisfying to reach a specified target accuracy. The advantage comes instead from the fact that it is possible to more rapidly, i.e. computationally more efficiently, reach the specified target by using fewer steps.

As shown in Table 6.2, the global error in the Euler method is $O(h)$. This means that if the step size is reduced by a factor of $1/2$, the overall error will be reduced by a factor of $1/2$. In contrast, the global error in the fourth-order Runge–Kutta method is $O(h^4)$, which means that the error will decrease by a factor of $1/16$ when the step size is reduced by a factor of $1/2$.

Why not use higher-order methods beyond the fourth order? One reason is that the Runge–Kutta methods with higher order use more function evaluations, which means that the number of computations increases. The objective of numerical solutions is to obtain a solution with sufficient accuracy, not an exact solution. Consequently, it is often not necessary to use higher-order methods because the improvement in accuracy is offset by the increase in computational effort. A better strategy for increasing accuracy is instead to use adaptive step size methods.

6.1.4   Adaptive step size methods and error control

Using a small fixed step size to resolve fast changes in a solution to maintain good accuracy could mean that the rest of the solution is solved very slowly. Often the best strategy for obtaining high accuracy, and at the same time minimizing computational effort, is to use adaptive step size methods when solving IVPs. This means that the step size is controlled and locally adjusted automatically during the solution where necessary. Since the basic idea with the adaptive methods is to determine whether a proper step size h is being used, an estimate of the local error is required.

Instead of evaluating an error by solving the IVP twice, by using two different fixed step sizes, e.g. h and $h/2$, the adaptive methods rely on error estimations from embedded pairs of different orders. As the solution proceeds, two solutions with different orders can be calculated in parallel and compared at each step to ensure that the accuracy is within acceptable tolerance. The simplest way to do this is to compare the error estimate ε_i, or the relative error estimate ε_i/y_i, with the error tolerance, and then double or halve the step size, depending on the current error. If the error exceeds the tolerance, the step must be repeated, using a reduced step size to improve accuracy. In regions where solution accuracy is met unnecessarily well, the step size can be increased to reduce computational effort.

The Runge–Kutta–Fehlberg method is one example of an adaptive step size method. It compares the solutions from a fourth-order Runge–Kutta method with a solution from a fifth-order Runge–Kutta method to determine the optimal step size. The optimal step, h_{opt}, is determined by evaluating the difference between the fourth- and fifth-order Runge–Kutta solutions, and accounting for the specified error tolerance.

Another method that uses fourth- and fifth-order embedded pairs is the Dormand–Prince method. The Dormand–Prince method is more accurate than the Runge–Kutta–Fehlberg method and it is used by the MATLAB ode45 solver. Both methods have in common that the difference between the fourth- and fifth-order accurate solutions is calculated to determine the error, and to adapt the step size. The error estimate, ε_{n+1}, for the step is

$$\varepsilon_{n+1} = |z_{n+1} - y_{n+1}|, \tag{6.16}$$

where the fifth-order accurate solution, z_{n+1}, is

$$z_{n+1} = y_n + h\left(\frac{35}{384}k_1 + \frac{500}{1113}k_3 + \frac{125}{192}k_4 - \frac{2187}{6784}k_5 + \frac{11}{84}k_6\right). \tag{6.17}$$

Note that k_2 is not used except for calculating k_3, and so on. The definitions are

$$k_1 = f(x_n, y_n),$$

$$k_2 = f\left(x_n + \frac{1}{5}h, y_n + h\frac{1}{5}k_1\right),$$

$$k_3 = f\left(x_n + \frac{3}{10}h, y_n + h\left(\frac{3}{40}k_1 + \frac{9}{40}k_2\right)\right),$$

$$k_4 = f\left(x_n + \frac{4}{5}h, y_n + h\left(\frac{44}{45}k_1 - \frac{56}{15}k_2 + \frac{32}{9}k_3\right)\right),$$

$$k_5 = f\left(x_n + \frac{8}{9}h, y_n + h\left(\frac{19\,372}{6561}k_1 - \frac{25\,360}{2187}k_2 + \frac{64\,448}{6561}k_3 - \frac{212}{729}k_4\right)\right),$$

$$k_6 = f\left(x_n + h, y_n + h\left(\frac{9017}{3168}k_1 - \frac{355}{33}k_2 + \frac{46\,732}{5247}k_3 + \frac{49}{176}k_4 - \frac{5103}{18\,656}k_5\right)\right),$$

and the fourth-order accurate solution, y_{n+1}, is

$$y_{n+1} = y_n + h\left(\frac{5179}{57\,600}k_1 + \frac{7571}{16\,695}k_3 + \frac{393}{640}k_4 - \frac{92\,097}{339\,200}k_5 + \frac{187}{2100}k_6 + \frac{1}{40}k_7\right), \tag{6.18}$$

where

$$k_7 = f(x_n + h, z_{n+1}).$$

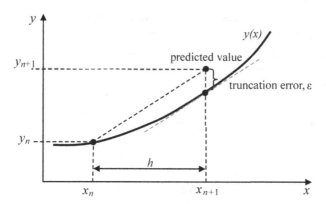

Figure 6.8. Principle of the implicit Euler method.

Consequently, the error estimate in the Dormand–Prince method is given by

$$\varepsilon_{n+1} = |z_{n+1} - y_{n+1}| = h \left| \frac{71}{57\,600}k_1 - \frac{71}{16\,695}k_3 + \frac{71}{1920}k_4 - \frac{17\,253}{339\,200}k_5 \right.$$
$$\left. + \frac{22}{525}k_6 - \frac{1}{40}k_7 \right|. \tag{6.19}$$

Given this estimate, ε_{n+1}, and a specified target tolerance for the discretization error, the adaptive method will decrease the step size to comply with the tolerance. If the accuracy is met unnecessarily well, the step size can be increased to reduce the computational effort. Furthermore, the step size can be controlled or bounded by a specified minimum step size, h_{min}, and maximum step size, h_{max}, to improve stability.

6.1.5   Implicit methods and stability

All the methods presented so far, e.g. the Euler and the Runge–Kutta methods, are examples of explicit methods, as the numerical solution at y_{n+1} has an explicit formula. Explicit methods, however, have problems with stability, and there are certain stability constraints that prevent the explicit methods from taking very large time steps. Stability analysis can be used to show that the explicit Euler method is conditionally stable, i.e. the step size has to be chosen sufficiently small to ensure stability. This conditional stability, i.e. the existence of a critical step size beyond which numerical instabilities manifest, is typical for all explicit methods. In contrast, the implicit methods have much better stability properties. Let us introduce the implicit backward Euler method,

$$y_{n+1} = y_n + hf(x_{n+1}, y_{n+1}). \tag{6.20}$$

In contrast to the forward (explicit) Euler method, which uses the slope at the left-hand side to step across the interval, the implicit version of the Euler method crosses the interval by using the slope at the right-hand side, as shown in Figure 6.8. The implicit formula does not give any direct approximation of y_{n+1}, instead an iterative method, e.g. the Newton method, is added inside the loop, thus advancing the differential equation to solve for y_{n+1}. This obviously comes at the price of more computation, but allows stability

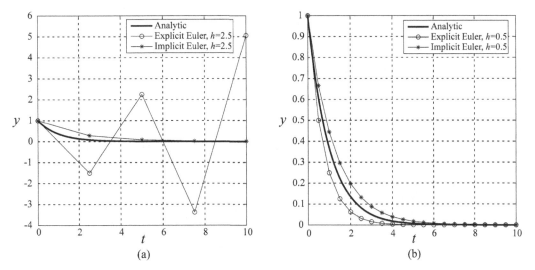

Figure 6.9. Comparison between the explicit and implicit Euler methods. (a) Step size $h = 2.5$; (b) step size $h = 0.5$.

also with comparatively large step size. This makes the implicit methods efficient and useful. The implicit Euler method is unconditionally stable for $h > 0$. Without using a rigorous stability analysis, the issue with stability is shown in Example 6.2.

Example 6.2 Numerical stability

Consider the problem of prediciting the concentration of a reactant in an isothermal batch reactor, assuming a first-order reaction, where the reaction rate, r, is given by the product of the reaction rate constant, k, and the concentration of the reactant, y:

$$\begin{cases} \dfrac{dy}{dt} = f(t, y) = -r = -ky = -y/\tau, \\ y(0) = 1, \\ t \in [0, 10]. \end{cases}$$

In this equation, we have introduced the time constant of the system, which is the inverse of the rate constant k, i.e. $\tau = 1/k$. This simple problem involves a linear differential equation that allows us to investigate how the implicit and the explicit methods behave as the step size is modified. Recall that the explicit Euler method is given by $y_{n+1} = y_n + hf(x_n, y_n)$. This means that

$$y_{n+1} = y_n - h\frac{y_n}{\tau} = y_n\left(1 - \frac{h}{\tau}\right). \tag{6.21}$$

Therefore, if the time step, h, is larger than the time constant, the explicit method will predict negative concentrations and easily diverge, as shown in Figure 6.9(a). In contrast, the implicit Euler method $y_{n+1} = y_n + hf(x_{n+1}, y_{n+1})$ has the advantage of stability,

$$y_{n+1} = y_n - h\frac{y_{n+1}}{\tau}, \tag{6.22}$$

which can be written as

$$y_{n+1} = \frac{y_n}{1 + h/\tau}. \tag{6.23}$$

Consequently, the implicit method will be stable even for the poor choice of time step h. However, in order to achieve an appropriate accuracy, the step size h has to be chosen reasonably small. This simple example differs from most practical applications in one very important aspect. In this case, it was simple, using algebra, to rewrite the original implicit formula as an explicit one for evaluation. This is usually not the case in practical applications. In such cases, the implicit method is more complex to use, and it involves solving for y_{n+1}, using indirect means.

Let us now compare the explicit and implicit Euler techniques using two different step sizes, $h = 2.5$ and $h = 0.5$, for the interval $t \in [0, 10]$, using $\tau = 1$ and the initial value $y_0 = 1$. From what we have just learned, the stability issues are manifested as divergence in the explicit solver when too large time steps are used. As shown in Figure 6.9(a), on using a step size larger than the time constant of the system, the explicit method diverges, whereas the implicit formula does not. The solutions are comparable for smaller time steps (Figure 6.9(b)). Note that both these methods have low accuracy due to their low order.

Example 6.3 Implict Euler–Newton method

Consider the problem of predicting the concentration of a reactant in an isothermal batch reactor, where the reaction kinetics is second order (non-linear problem). The reaction rate, r, is given by $r = ky^2$, where k is the reaction rate constant ($k = 1$) and y is the concentration of the reactant. We have

$$\begin{cases} \dfrac{dy}{dt} = f(t, y) = -r = -y^2, \\ y(0) = 1, \\ t \in [0, 1]. \end{cases}$$

By using the the the implicit Euler method, $y_{n+1} = y_n + hf(x_{n+1}, y_{n+1})$, and a time step of $h = 0.1$, we obtain, at the first time step,

$$y_1 = y_0 - hy_1^2.$$

As the value y_1 is unknown we need to iterate, e.g. use the Newton method, to find the solution. Let us denote $y_1 = z$ and implement the Newton method to determine the solution at the first time step, $t = 0.1$. First we reformulate the non-linear equation to $F(z) = 0$; this gives

$$F(z) = z - y_0 + hz^2 = z - 1 + 0.1z^2 = 0.$$

The Newton method to find a root of $F(z) = 0$ and with an initial guess, z_i, is given by

$$z_{i+1} = z_i - \frac{F(z_i)}{F'(z_i)} \quad \text{for} \quad i = 0, 1, 2 \ldots$$

We can obtain an initial guess, z_0, by using the forward Euler method, $y_1 = y_0 - hy_0^2 = 1 - 0.1 \cdot 1^2 = 0.9$. The first iteration with the Newton method gives

$$z_1 = z_0 - \frac{F(z_0)}{F'(z_0)} = z_0 - \frac{z_0 - 1 + 0.1z_0^2}{1 + 0.2z_0} = 0.9 - \frac{0.9 - 1 + 0.1 \cdot 0.9^2}{1 + 0.2 \cdot 0.9}$$
$$= 0.916101694915254,$$

and the second iteration gives

$$z_2 = z_1 - \frac{F(z_1)}{F'(z_1)} = z_1 - \frac{z_1 - 1 + 0.1z_1^2}{1 + 0.2z_1} = 0.916079783140194.$$

Finally the solution converges at $z = 0.916079783099616$, which is the reactant concentration, y_1, predicted by the implicit Euler method at the first time step $t = 0.1$. This procedure is repeated for the following nine time steps to determine the final reactant concentration. To sum up, the implicit Euler method involves more computation; it does not improve accuracy because it is only first-order accurate, but it significantly improves stability.

6.1.6 Multistep methods and predictor–corrector pairs

In contrast to the one-step methods, e.g. the Runge–Kutta method, where the solution at x_{n+1} is calculated from the solution at x_n, multistep methods calculate the solution from previous steps, e.g. x_n, x_{n-1}, x_{n-2}. In other words, multistep methods attempt to gain efficiency by keeping and using the information from previous steps rather than discarding it. Therefore, these methods are more efficient than the single-step Runge–Kutta methods.

6.1.6.1 Explicit multistep methods

Among the explicit multistep methods, the Adams–Bashforth methods are the most widely used. The second-order (global error) Adams–Bashforth two-step method is

$$y_{n+1} = y_n + h \left(\frac{3}{2} f(x_n, y_n) - \frac{1}{2} f(x_{n-1}, y_{n-1}) \right). \tag{6.24}$$

The fourth-order, four-step Adams–Bashforth method,

$$y_{n+1} = y_n + h \left(\frac{55}{24} f(x_n, y_n) - \frac{59}{24} f(x_{n-1}, y_{n-1}) + \frac{37}{24} f(x_{n-2}, y_{n-2}) \right.$$
$$\left. - \frac{9}{24} f(x_{n-3}, y_{n-3}) \right), \tag{6.25}$$

requires information about $f(x_n, y_n)$, $f(x_{n-1}, y_{n-1})$, $f(x_{n-2}, y_{n-2})$, and $f(x_{n-3}, y_{n-3})$; however, only $f(x_n, y_n)$ needs to be calculated in this step as the other values have been calculated in previous steps. Obviously the multistep methods need help getting started because they require information about previous steps. A one-step method is often used for this start-up phase. Compared to the fourth-order Runge–Kutta method, which needs four function evaluations, this method requires only one function evaluation after the start-up phase. The advantage is clear; the multistep method can be used to obtain the same accuracy with less computational effort.

6.1.6.2   Implicit multistep methods

Neither the Runge–Kutta nor the Adams–Bashforth methods can handle stiff differential equations well. The Adams–Moulton method is an implicit multistep method that can handle stiff problems better (stiff problems are dicussed later in this chapter). The two-step Adams–Moulton method (third-order accurate) is

$$y_{n+1} = y_n + h \left(\frac{5}{12} f(x_{n+1}, y_{n+1}) + \frac{8}{12} f(x_n, y_n) - \frac{1}{12} f(x_{n-1}, y_{n-1}) \right), \quad (6.26)$$

and the three-step Adams–Moulton method (fourth-order accurate) is

$$y_{n+1} = y_n + h \left(\frac{9}{24} f(x_{n+1}, y_{n+1}) + \frac{19}{24} f(x_n, y_n) - \frac{5}{24} f(x_{n-1}, y_{n-1}) \right.$$
$$\left. + \frac{1}{24} f(x_{n-2}, y_{n-2}) \right). \quad (6.27)$$

The implicit multistep methods add stability but require more computation to evaluate the implicit part. In addition, the error coefficient of the Adams–Moulton method of order k is smaller than that of the Adams–Bashforth method of the same order. As a consequence, the implicit methods should give improved accuracy. In fact, the error coefficient for the implicit fourth-order Adams–Moulton method is 19/720, and for the explicit fourth-order Adams–Bashforth method it is 251/720. The difference is thus about an order of magnitude. Pairs of explicit and implicit multistep methods of the same order are therefore often used as predictor–corrector pairs. In this case, the explicit method is used to calculate the solution, $\tilde{y}_{n+1}$, at x_{n+1}. Furthermore, the implicit method (corrector) uses $\tilde{y}_{n+1}$ to calculate $f(x_{n+1}, \tilde{y}_{n+1})$, which replaces $f(x_{n+1}, y_{n+1})$. This allows the solution, y_{n+1}, to be improved using the implicit method. The combination of the Adams–Bashforth and the Adams–Moulton methods as predictor–corrector pairs is implemented in some ODE solvers. The MATLAB ode113 solver is an example of a variable-order Adams–Bashforth–Moulton multistep solver.

6.1.7   Systems of ODEs

Systems of coupled ODEs might arise from reformulating a higher-order differential equation to a system of first-order differential equations, or as a description of a system that consists of coupled variables. Systems of differential equations can be solved as an extension of the methodology for a single differential equation. The principle is shown in Example 6.4, which considers a stirred tank reactor.

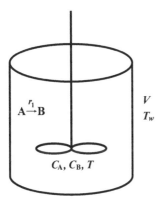

Figure 6.10. Batch reactor with perfect mixing.

Example 6.4 System of ODEs

Consider an exothermic reaction occurring in a batch reactor, as shown in Figure 6.10. The reaction mechanism is a first-order reaction, $A \overset{r_1}{\to} B$, and the Arrhenius reaction rate law applies, i.e.

$$r_1 = kC_A = k_0 e^{-E_a/RT} C_A. \tag{6.28}$$

The system is described by the material balance,

$$\frac{dC_A}{dt} = -r_1 = -k_0 e^{-E_a/RT} C_A = f_1(C_A, T), \tag{6.29}$$

and the energy balance,

$$\frac{dT}{dt} = \frac{r_1(-\Delta H)V - UA(T - T_w)}{V\rho c_p} = f_2(C_A, T), \tag{6.30}$$

with the initial conditions $C_A(0) = C_{A0}$, $T(0) = T_0$.

Here, V is the reactor volume, $UA(T - T_w)$ is the external cooling for controlling the reaction temperature, and $r_1(-\Delta H)V$ is the heat caused by chemical reaction. As these equations are coupled, they must be advanced simultaneously. For simplicity, and for pedagogical reasons, we will use the explicit Euler method to solve this problem:

$$C_{A,n+1} = C_{A,n} + h f_1(C_{A,n}, T_n) = C_{A,n} - h k_0 e^{-E_a/RT_n} C_{A,n}, \tag{6.31}$$

$$T_{n+1} = T_n + h f_2(C_{A,n}, T_n) = T_n + \frac{h(k_0 e^{-E_a/RT_n} C_{A,n}(-\Delta H)V - UA(T_n - T_w))}{V\rho c_p}. \tag{6.32}$$

Solving Equations (6.31) and (6.32) obviously requires that the slopes $f_1(C_{A,n}, T_n)$ and $f_2(C_{A,n}, T_n)$ are calculated before C_A and T are updated. Using a fourth-order Runge–Kutta algorithm means that eight slopes need to be calculated in each step.

Table 6.3. Transformation of higher-order derivatives

Define y_i	ODE for y_i
$y_1 = y$	$\dfrac{dy_1}{dx} = y_2$
$y_2 = \dfrac{dy}{dx}$	$\dfrac{dy_2}{dx} = y_3$
$y_3 = \dfrac{d^2 y}{dx^2}$	$\dfrac{dy_3}{dx} = y_4$
$y_n = \dfrac{d^{n-1} y}{dx^{n-1}}$	$\dfrac{dy_n}{dx} = f_1(x, y, \dfrac{dy}{dx}, \dfrac{d^2 y}{dx^2}, \ldots, \dfrac{d^{n-1} y}{dx^{n-1}})$

6.1.8   Transforming higher-order ODEs

The methods described earlier in this chapter apply to higher-order differential equations, i.e.

$$\frac{d^n y}{dx^n} = f_1 \left(x, y, \frac{dy}{dx}, \frac{d^2 y}{dx^2}, \ldots, \frac{d^{n-1} y}{dx^{n-1}} \right). \tag{6.33}$$

Higher-order ODEs can be rewritten to a system of first-order equations. In this case, the numerical methods developed for solving single first-order ODEs can be extended directly to ODE systems. This means that the numerical methods for solving higher-order differential equations are reduced to integrating a system of first-order differential equations. The general principle for transforming a higher-order differential equation, Equation (6.33), is shown in Example 6.5 and is summarized in Table 6.3.

Example 6.5 Transformation

In this example we will consider the transformation of a second-order ODE to a system of two first-order ODEs. The second-order differential equation is given by

$$a(x)\frac{d^2 y}{dx^2} + b(x)\frac{dy}{dx} = c(x). \tag{6.34}$$

Let

$$y(x) = y_1(x) \tag{6.35}$$

and

$$\frac{dy_1}{dx} = y_2(x). \tag{6.36}$$

Subsequently, the second-order equation is transformed to the system

$$\begin{cases} \dfrac{dy_1}{dx} = y_2(x), \\ \dfrac{dy_2}{dx} = \dfrac{c(x) - b(x)y_2(x)}{a(x)}. \end{cases} \tag{6.37}$$

This system of first-order differential equations is integrated to solve the second-order ODE Equation (6.34), using any of the methods described in the previous sections.

6.1.9 Stiffness of ODEs

Differential equations must also be characterized according to their numerical stiffness, because only a few numerical methods can handle stiff problems. Without trying to give a precise mathematical definition of stiffness we will consider a problem as stiff when there are multiple time scales present that cause abrupt changes in a data series. This often occurs due to the separation of time scales, which is very common in models of chemical engineering systems. Unfortunately, the performance of the explicit numerical algorithms presented in this chapter, e.g. the Runge–Kutta methods, are not good because a faster time scale restricts how large-step size can be used without divergence, i.e. to maintain numerical stability. Consequently, the problem is considered stiff when the numerical solution has its step size limited more severely by the stability of the method than by its accuracy. In other words, the major problem with a non-stiff solver is that the computation time for stiff equation systems becomes very high. Whilst a non-stiff solver would require very small time steps throughout an entire interval, a stiff solver would require only small steps when the solutions rapidly change, i.e. to resolve step gradients in the interval.

The most popular methods for solving stiff problems are the backward differentiation formulas (BDFs) and the numerical differentiation formulas (NDFs). The MATLAB ode15s uses NDFs. These methods belong to the family of implicit methods and have good stability properties. The BDFs approximate the derivative of a function by using information from already computed values; they are methods that give an approximation to a derivative of a variable at x_n in terms of its function values $y(x)$ at x_n and at earlier locations. This means that BDFs are implicit, just like the Adams–Moulton methods. Compared to the Adams–Moulton methods, however, BDFs are not as accurate for formulas of the same order. Furthermore, they are not stable for orders of 7 and higher. However, at the orders at which the formulas are stable, they are much more stable than the Adams–Moulton methods, which explains why they are often used to solve stiff problems. The simplest BDF is a one-step formula

$$y_{n+1} = y_n + hf(x_{n+1}, y_{n+1}), \tag{6.38}$$

which is the backward Euler method. The higher-order BDFs involve previously computed solution values, for example, a second-order BDF,

$$y_{n+1} = \frac{1}{3}(4y_n - y_{n-1} + 2hf(x_{n+1}, y_{n+1})). \tag{6.39}$$

General purpose methods adapt the step size to the solution of the problem, and they also adapt the order of the method, which allows both high accuracy and stability. The stability of the methods generally becomes worse as the order increases. The maximum order can be specified for stability reasons. A simple numerical experiment can be used to show how the selection of ODE solver affects a solution. Consider Example 6.6, which becomes stiff for large α values.

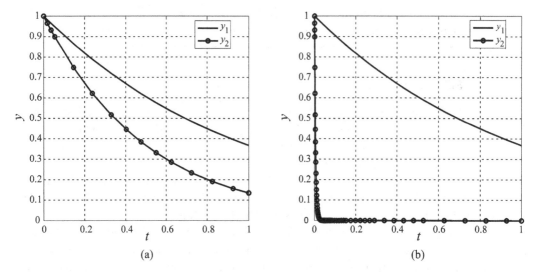

Figure 6.11. Solution to (a) a non-stiff system, $\alpha = 2$, (b) a stiff system, $\alpha = 200\,000$.

Example 6.6 Stiff differential equation system

Consider the potentially stiff system

$$\frac{dy_1}{dt} = -y_1(t),$$

$$\frac{dy_2}{dt} = -\alpha y_2(t),$$

$$y_1(0) = 1,$$

$$y_2(0) = 1,$$

$$t \in [0, 1].$$

The analytical solution to this problem is given by

$$\begin{cases} y_1(t) = e^{-t}, \\ y_2(t) = e^{-\alpha t}. \end{cases}$$

For large α values, the system becomes stiff as it contains components that vary with different speeds, i.e. $y_2(t)$ approaches zero much faster than $y_1(t)$ does. The solutions to this problem for $\alpha = 2$ and $\alpha = 200\,000$ are shown in Figure 6.11. By analyzing the number of steps required by the different methods, it becomes clear how the stiff solvers outperform the explicit methods. Calculations have been performed for different values of α with different error tolerances; the results are summarized in Table 6.4.

We can draw several conclusions from Table 6.4 after what we have learned in this chapter. We can conclude that for non-stiff problems and moderate error tolerances all methods work efficiently. When the error tolerance is stricter, the higher-order method ode45 performs better than the ode23 method.

Table 6.4. Number of steps used by the different MATLAB solvers

| | Relative tolerance = 10^{-3} | | | Relative tolerance = 10^{-9} | | |
| | Absolute tolerance = 10^{-6} | | | Absolute tolerance = 10^{-12} | | |
	$\alpha = 2$	$\alpha = 200$	$\alpha = 200\,000$	$\alpha = 2$	$\alpha = 200$	$\alpha = 200\,000$
ode23	11	105	79 620	688	3708	82 985
ode45	10	73	60 274	39	285	60 486
ode23s	11	53	58	858	4627	4686
ode15s	14	75	91	83	438	461

For stiff problems, both of the explicit methods (ode23 and ode45) perform poorly, regardless of the error tolerance specified. Both require many steps to advance through the interval $t \in [0, 1]$. As mentioned before, these two explicit solvers use adaptive step sizes, but there are also stability constraints that prevent them from taking large time steps, even if the problem would seem to allow for this in regions where the derivative is low. In contrast, the implicit ODE solvers have much better stability and solve the problem more efficiently. When the error tolerence is very strict, ode15s performs better than ode23s; this is because it is a variable-order method, of orders 1–5, whereas ode23s is based on a formulation of orders of 2 and 3. More information about these numerical algorithms can be found at the end of this chapter.

6.2 Boundary-value problems

A differential equation that has data given at more than one value of the independent variable is a boundary-value problem (BVP). Consequently, the differential equation must be of at least second order. The solution methods for BVPs are different compared to the methods used for initial-value problems (IVPs). An overview of a few of these methods will be presented in Sections 6.2.1–6.2.3. The shooting method is the first method presented. It actually allows initial-value methods to be used, in that it transforms a BVP to an IVP, and finds the solution for the IVP. The lack of boundary conditions at the beginning of the interval requires several IVPs to be solved before the solution converges with the BVP solution. Another method presented later on is the finite difference method, which solves the BVP by converting the differential equation and the boundary conditions to a system of linear or non-linear equations. Finally, the collocation and finite element methods, which solve the BVP by approximating the solution in terms of basis functions, are presented.

6.2.1 Shooting method

The shooting method uses methods similar to the ones used for solving IVPs; the main difference comes from the iterative approach that is required for solving BVPs. The

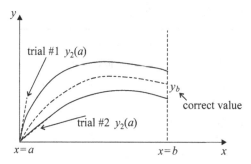

Figure 6.12. Principle of the shooting method.

iterative approach is needed to determine the missing initial values that are consistent with the boundary values. Consider the following second-order BVP:

$$\begin{cases} \dfrac{d^2y}{dx^2} = f\left(x, y, \dfrac{dy}{dx}\right), \\ y(a) = y_a, \\ y(b) = y_b, \\ x \in [a, b], \end{cases} \tag{6.40}$$

which can be transformed to a system of first-order differential equations using the method described in Section 6.1, i.e. by setting $y_1 = y$ and $y_2 = dy/dx$,

$$\begin{cases} \dfrac{dy_1}{dx} = y_2, \\ \dfrac{dy_2}{dx} = f(x, y_1, y_2). \end{cases}$$

In order to apply any of the IVP methods for solving this system of differential equations, we need one condition each for the variables y_1 and y_2. However, having two conditions for y_1 and none for the other variable does not help. Instead we need to guess $y_2(a)$ and perform the integration to the point $x = b$, as illustrated in Figure 6.12. For other BVPs, the derivative, $y_2(a)$, might be given and $y_1(a)$ might need to be guessed (as in Example 6.7).

In Figure 6.12 two different slopes have been guessed, resulting in two different solutions, neither of which match the boundary value y_b. Different root-finding algorithms may be used to find the missing initial condition, e.g. the secant method or the Newton method. When the missing initial condition has been found, the solution of the IVP at $y(b)$ will match y_b.

This iterative procedure is summarized in the following algorithm:

- guess $y_2(a)$ (or $y_1(a)$ if that is missing);
- integrate the system of ODEs over the entire domain, $x \in [a, b]$;
- terminate if $\varepsilon = |y_1(b) - y_b| < tol$; or else
- adjust $y_2(a)$ and continue to point 2.

Adjustment of the initial guess can be done using the secant method, which can be applied as soon as two trials have been completed,

$$y_{2,i+1} = y_{2,i} - \varepsilon_i \frac{y_{2,i} - y_{2,i-1}}{\varepsilon_i - \varepsilon_{i-1}}. \tag{6.41}$$

As an alternative, the Newton method can be used. The reader is referred to the literature to learn more about these methods.

Example 6.7 Shooting method

Reaction rates inside a spherical isothermal porous catalyst particle depend on the rate of diffusion and the kinetics. When the reaction is fast compared to the diffusion, e.g. at higher temperatures, the reactants will be consumed near the surface and not all the catalyst will be effectively used. The ratio of the actual reaction rate throughout the particle to the reaction rate without any diffusion limitation is a good measure of how effectively the catalyst is used. This is called the effectiveness factor, η. Consider calculating the effectiveness factor for a spherical catalyst particle for a first-order reaction at the conditions specified below. The governing equation and boundary conditions are given in the dimensionless form, and a first-order reaction is assumed:

$$\begin{cases} \dfrac{d^2C}{dr^2} + \dfrac{2}{r}\dfrac{dC}{dr} - \phi^2 C = 0, \\ C(1) = 1, \\ \dfrac{dC}{dr}\bigg|_{r=0} = 0, \\ r \in [0, 1]. \end{cases} \tag{6.42}$$

Here the dimensionless variables are concentration c, radius r, and $\phi = R\sqrt{k}/D_e$ is the Thiele modulus, a dimensionless number used to describe the relation between the reaction and diffusion rates. In this ODE-BVP, the first boundary condition specifies the concentration at the catalyst surface, and the second boundary condition specifies no flux at the center of the particle. In order to use the shooting method, the BVP is rewritten using $dC/dr = w$, which gives a system of first-order ODEs,

$$\frac{dw}{dr} = \phi^2 C - \frac{2}{r}w,$$
$$\frac{dC}{dr} = w. \tag{6.43}$$

At this point, ordinary ODE-IVP solvers can be used to solve the problem, but, due to the lack of boundary conditions at the beginning of the interval, i.e. at $r = 0$, the $C(r = 0)$ value needs to be guessed. When a converged solution is obtained, i.e. when $C(r = 1) = 1$, the effectiveness factor can be calculated using

$$\eta = \frac{\int_0^1 C(r)r^2\,dr}{\int_0^1 C|_{r=1}r^2\,dr}.$$

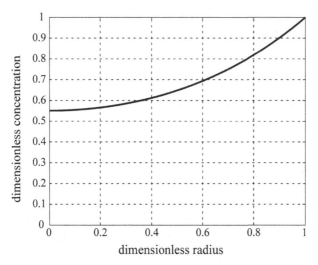

Figure 6.13. Concentration profile in the catalyst particle of Example 6.7 for $\phi = 2$.

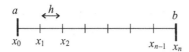

Figure 6.14. Computational grid, with n grid points.

When $\phi = 2$ the solution is obtained by guessing $C(r = 0) = 0.55$. The concentration profile in the catalyst for $\phi = 2$ is shown in Figure 6.13, and the effectiveness factor, $\eta = 0.81$, agrees with the analytical value that can be determined for this first-order kinetics.

6.2.2 Finite difference method for BVPs

The basic principle in using the finite difference method to solve BVPs is to replace all the derivatives in the differential equation with difference-quotient approximations. First, the interval $a < x < b$ is discretized into n equally spaced intervals (an unequal spaced interval may also be used):

$$a = x_0 < x_1 < x_2 < x_3 < \cdots < x_n = b.$$

Consequently, the mesh (or grid) points are given by

$$x_i = a + i \times h, \text{ for } i = 0, 1, 2, \ldots, n,$$

where $h = (b - a)/n$ is the mesh size, as shown in Figure 6.14.

Assume a general linear BVP

$$\frac{d^2 y}{dx^2} = p(x)\frac{dy}{dx} + q(x)y + r(x),$$
$$y(a) = y_a,$$
$$y(b) = y_b. \tag{6.44}$$

From the Taylor theorem, we can derive the difference approximation for the first- and second-order derivatives by expanding y about x_i. When using central differences, the first-order derivative is approximated by

$$\frac{dy}{dx}\bigg|_{x_i} = \frac{y(x_{i+1}) - y(x_{i-1})}{2h} + O(h^2), \tag{6.45}$$

which means that the approximation is $O(h^2)$. The second-order derivative (using central difference approximations) is approximated by

$$\frac{d^2y}{dx^2}\bigg|_{x_i} = \frac{y(x_{i+1}) - 2y(x_i) + y(x_{i-1})}{h^2} + O(h^2). \tag{6.46}$$

By substituting Equations (6.45) and (6.46) into Equation (6.44), and introducing $p_i = p(x_i)$, $q_i = q(x_i)$, $r_i = r(x_i)$, the discretized version of Equation (6.44) is obtained as follows:

$$\frac{y(x_{i+1}) - 2y(x_i) + y(x_{i-1})}{h^2} = p_i \frac{y(x_{i+1}) - y(x_{i-1})}{2h} + q_i y(x_i) + r_i,$$
$$\text{for } i = 1, 2, \ldots, n-1, \tag{6.47}$$

which simplifies to

$$(1 + hp_i/2)y(x_{i-1}) - (2 + h^2 q_i)y(x_i) + (1 - hp_i/2)y(x_{i+1}) = h^2 r_i. \tag{6.48}$$

This means there are $n-1$ unknowns, i.e. $y_1, y_2, \ldots, y_{n-1}$, and $n-1$ algebraic equations that should be solved for. This can be written in the matrix form, $Ay = f$, where

$$A = \begin{bmatrix} -2 - h^2 q_1 & 1 - hp_1/2 & 0 & 0 & . & . \\ 1 + hp_2/2 & -2 - h^2 q_2 & 1 - hp_2/2 & 0 & 0 & . \\ 0 & & & & & \\ 0 & & & & & \\ . & & 0 & 1 + hp_{n-2}/2 & -2 - h^2 q_{n-2} & 1 - hp_{n-2}/2 \\ . & & 0 & 0 & 1 + hp_{n-1}/2 & -2 - h^2 q_{n-1} \end{bmatrix}, \tag{6.49}$$

$$y = \begin{bmatrix} y_1 \\ y_2 \\ . \\ . \\ . \\ y_{n-2} \\ y_{n-1} \end{bmatrix}, \quad \text{and} \quad f = \begin{bmatrix} h^2 r_1 - (1 + hp_1/2)y_a \\ h^2 r_2 \\ . \\ . \\ . \\ h^2 r_{n-2} \\ h^2 r_{n-1} - (1 - hp_{n-1}/2)y_b \end{bmatrix}. \tag{6.50}$$

Consequently, the approximative solution to the BVP is calculated on a computational mesh, and it results in a system of algebraic equations. In this example, the problem is linear and the Gaussian elimination can be used to solve the equation system ($n-1$ algebraic equations). Dirichlet boundary conditions were specified in this problem. Note that the boundary conditions are accounted for in two of these equations, i.e.

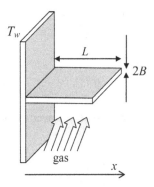

Figure 6.15. Cooling-fin design.

the first and last rows in the matrix Equations (6.50). The finite difference method can be summarized as follows.

(1) Convert the continuous variable to discrete variables. The mesh for the independent variable may be evenly or unevenly spaced.
(2) Use the difference approximation to replace the derivatives.
(3) Implement the boundary conditions.
(4) Solve the equation system using direct or indirect methods.

As shown here, the implementation of Dirichlet boundary conditions is straightforward. In contrast, if the problem contains a Neumann boundary condition, i.e. a derivative on the boundary, a little more work will be needed.

If the BVP contains a Neumann boundary condition, one way to solve the problem is to extend the original computational domain, i.e. from nh to $(n + 1)h$. This means introducing a fictitious boundary. This obviously requires one more grid point outside the original grid; the additional one is at x_{n+1}, as shown in Example 6.8. By introducing the fictitious node, the accuracy $O(h^2)$ is maintained.

Example 6.8 Finite difference method

Cooling fins, Figure 6.15, are frequently used to improve heat transfer from walls to gases, because gases have low thermal conductivity. The fin efficiency factor, η, is defined as the ratio between the actual heat transfer and the heat transfer if the entire fin has the wall temperature T_w. The temperature obviously depends on fin dimensions (length and thickness) and on the thermal conductivity of the fin material. In order to quantify the efficiency of a fin, the temperature profile along the fin must be determined.

The heat transfer is governed by the following differential equation:

$$\frac{d^2 T}{dx^2} = \frac{\alpha}{\lambda B}(T - T_{amb}), \tag{6.51}$$

where α is the heat transfer coefficient and λ is the thermal conductivity. This is a second-order differential equation that requires two boundary conditions. We assume that the temperature at $x = 0$ equals T_w, i.e.

$$T(0) = T_w, \tag{6.52}$$

and that the flux at the tip is negligible, i.e.

$$\frac{\partial T}{\partial x}\bigg|_{x=L} = 0. \tag{6.53}$$

By introducing the dimensionless variables

$$\theta = \frac{T - T_{amb}}{T_w - T_{amb}}, \quad \xi = x/L, \quad H = \sqrt{\frac{\alpha L^2}{\lambda B}},$$

Equation (6.51) can be written as

$$\frac{d^2\theta}{d\xi^2} = H^2\theta, \tag{6.54}$$

and the new boundary conditions become

$$\theta(0) = 1, \tag{6.55}$$

$$\frac{d\theta}{d\xi}\bigg|_{\xi=1} = 0. \tag{6.56}$$

By using the central difference approximation for the second-order derivative, the discretized version of Equation (6.54) is obtained:

$$\frac{\theta(\xi_{i+1}) - 2\theta(\xi_i) + \theta(\xi_{i-1})}{h^2} - H^2\theta(\xi_i) = 0, \tag{6.57}$$

which simplifies to

$$\theta(\xi_{i-1}) - (2 + h^2 H^2)\theta(\xi_i) + \theta(\xi_{i+1}) = 0. \tag{6.58}$$

For simplicity, the domain $\xi \in [0, 1]$ is discretized into four equally spaced intervals, $h = (1 - 0)/4 = 0.25$, and we solve the problem for $H = 1$. This results in the linear equation system

$$\begin{aligned}
\theta_0 - (2 + h^2)\theta_1 + \theta_2 &= 0, \\
\theta_1 - (2 + h^2)\theta_2 + \theta_3 &= 0, \\
\theta_2 - (2 + h^2)\theta_3 + \theta_4 &= 0, \\
\theta_3 - (2 + h^2)\theta_4 + \theta_5 &= 0.
\end{aligned} \tag{6.59}$$

Note that a Dirichlet boundary condition is specified for the left-hand side of the domain, i.e. $\theta_0 = 1$. In order to implement the Neumann boundary condition on the right-hand side of the domain (the no-heat-flux condition on the tip), we can extend the computational domain from $\xi_i = i \times h$ for $i = 0, 1, 2, \ldots, n$ to $i = 0, 1, 2, \ldots, n + 1$. For this reason, the dependent variable is also calculated at the fictitious boundary outside the domain, at ξ_5. The Neumann boundary condition, i.e. the derivative at the

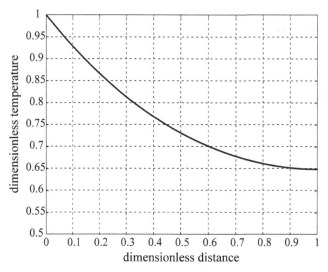

Figure 6.16. Temperature profile in the cooling fin of Example 6.8 for $H = 1$, grid resolution $h = 0.1$.

right-hand side of the domain, is approximated by the central difference formula to maintain the second-order accuracy of the method,

$$\frac{d\theta}{d\xi}\bigg|_{\xi_4} \approx \frac{\theta_5 - \theta_3}{2h} = 0, \qquad (6.60)$$

which yields $\theta_5 - \theta_3 = 0$. In total we have five equations and five unknowns in the equation system Equations (6.59) and (6.60), which can be written as follows:

$$\begin{bmatrix} -(2+h^2) & 1 & 0 & 0 & 0 \\ 1 & -(2+h^2) & 1 & 0 & 0 \\ 0 & 1 & -(2+h^2) & 1 & 0 \\ 0 & 0 & 1 & -(2+h^2) & 1 \\ 0 & 0 & -1 & 0 & 1 \end{bmatrix} \begin{bmatrix} \theta_1 \\ \theta_2 \\ \theta_3 \\ \theta_4 \\ \theta_5 \end{bmatrix} = \begin{bmatrix} -1 \\ 0 \\ 0 \\ 0 \\ 0 \end{bmatrix}. \qquad (6.61)$$

Equation (6.61) can be solved by using Gaussian elimination, and the predicted temperature profile inside the cooling fin is $\theta_1 = 0.8396$, $\theta_2 = 0.7318$, $\theta_3 = 0.6696$, $\theta_4 = 0.6493$ (and $\theta_5 = 0.6696$).

Note that θ_5 is the solution at the fictitious boundary (outside the domain), it can be excluded when plotting the solution on the domain of interest. In order to reduce the error, the domain is simply discretized into smaller intervals. The solution to the problem using finite differences and a mesh with $h = 0.1$ is shown in Figure 6.16.

The error, evaluated at $\xi = 0.5$, for different grid resolutions is shown in Figure 6.17. The error plotted in this log-log plot is a straight line with a slope equal to 2, which can be expected because the central difference approximation is second-order accurate, $O(h^2)$.

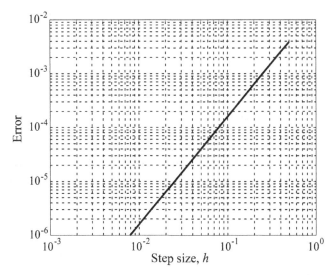

Figure 6.17. Error vs. step size, Example 6.8.

6.2.3 Collocation and finite element methods

Both the collocation and the finite element methods are function approximation methods. Similar to the finite difference method, the strategy here is to reduce the differential equation to a set of algebraic equations that can be solved. Instead of discretizing the differential equation by replacing the various derivatives with difference-quotient approximations, the solution is given a functional form,

$$y(x) = \sum a_i \phi_i(x). \tag{6.62}$$

In other words, the function approximation methods find a solution by assuming a particular type of function, a trial (basis) function, over an element or over the whole domain, which can be polynomial, trigonometric functions, splines, etc. These functions contain unknown parameters that are determined by substituting the trial function into the differential equation and its boundary conditions. In the collocation method, the trial function is forced to satisfy the boundary conditions and to satisfy the differential equation exactly at some discrete points distributed over the range of the independent variable, i.e. the residual is zero at these collocation points. In contrast, in the finite element method, the trial functions are defined over an element, and the elements are joined together to cover an entire domain.

The collocation method is a simple and effective method that is illustrated in the following example. Consider solving the following BVP problem by using the collocation method:

$$\begin{cases} \dfrac{d^2 y}{dx^2} = 6x, \\ y(0) = 0, \\ y(1) = 1, \\ x \in [0, 1]. \end{cases} \tag{6.63}$$

In this case, we will select a trial function, $\phi_i(x)$, which can fulfill the boundary conditions

$$y(x) = a_1 + a_2 x + a_3 x^2. \tag{6.64}$$

The derivatives of the polynomial, $y(x)$, are

$$\frac{dy}{dx} = a_2 + 2a_3 x \quad \text{and} \quad \frac{d^2 y}{dx^2} = 2a_3. \tag{6.65}$$

Three equations are obviously required to determine the unknowns a_1, a_2, a_3. By substituting Equation (6.65) into the differential equation Equation (6.63) and evaluating at certain mesh points, the problem is reduced to solving a system of equations in a_i. The collocation points should be selected at points that have a large influence on the function. We require that the BVP is satisfied in the middle of the domain, at the collocation point $x_2 = 0.5$, which yields

$$\frac{d^2 y}{dx^2} = 2a_3 = 6x_2 = 6 \cdot 0.5 = 3. \tag{6.66}$$

This means that $a_3 = 1.5$. The left boundary condition, $y(0) = 0$, gives

$$a_1 + a_2 \cdot 0 + a_3 \cdot 0^2 = 0,$$

and the right boundary condition, $y(1) = 1$, gives

$$a_1 + a_2 \cdot 1 + a_3 \cdot 1^2 = 1.$$

This gives the two other coefficients, $a_1 = 0$, $a_2 = -0.5$, and the polynomial becomes

$$y(x) = a_1 + a_2 x + a_3 x^2 = -0.5x + 1.5x^2. \tag{6.67}$$

The approximative solution to the BVP, i.e. the polynomial, and the true solution are shown in Figure 6.18.

6.3   Partial differential equations

Many phenomena in chemical engineering depend, in complex ways, on space and time, and often a mathematical model requires more than one independent variable to characterize the state of a system, i.e. the systems need to be described using partial differential equations, PDEs. Examples of such phenomena include chemical reactions, heat transfer, fluid flow, and population dynamics. For practical engineering applications, analytic solutions do not exist, and numerical methods need to be applied. This section is not intended to give a complete discussion of PDEs nor of solution methods. Instead, the aim is to introduce the terminology and some issues involved in solving PDEs. The discussion will be limited to linear PDEs that have two independent variables, e.g. space and time, or two space variables for a steady-state problem, with the form

$$A\frac{\partial^2 u}{\partial x^2} + B\frac{\partial^2 u}{\partial x \partial y} + C\frac{\partial^2 u}{\partial y^2} + f\left(\frac{\partial u}{\partial x}, \frac{\partial u}{\partial y}, u, x, y\right) = 0 \tag{6.68}$$

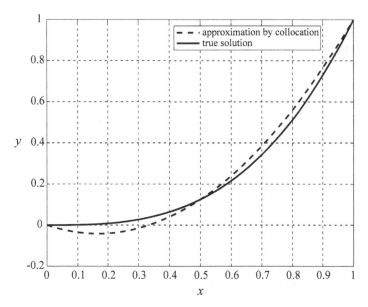

Figure 6.18. Solution of BVP using the collocation method.

or

$$Au_{xx} + Bu_{xy} + Cu_{yy} + f(u_x, u_y, u, x, y) = 0, \tag{6.69}$$

where the subscripts denote the partial derivatives of the dependent variable, u, and the focus is on the finite difference method.

6.3.1   Classification of PDEs

PDEs can be classified in different ways. The classification is important because the solution methods often apply only to a specified class of PDEs. To start with, PDEs can be classified by the number of variables, e.g. $u_t = u_{xx}$, which contain two independent variables, t and x. The order of a PDE is the order of the highest-order derivative that appears in the PDE. For example, $u_t = u_x$ is a first-order PDE, whereas $u_t = u_{xx}$ is a second-order PDE. In addition, it is important to make a distinction between non-linear and linear PDEs. An example of a well-known non-linear PDE is the Navier–Stokes equation, which describes the motion of fluids. In a linear PDE, the dependent variable and its derivatives appear in a linear fashion. The linear second-order PDE in Equation (6.69) can be classified as

  (i) parabolic, if $B^2 - 4AC = 0$;
 (ii) hyperbolic, if $B^2 - 4AC > 0$;
(iii) elliptic, if $B^2 - 4AC < 0$.

The $B^2 - 4AC$ term is referred to as the discriminant of the solution, and the behavior of the solution of Equation (6.69) depends on its sign. The smoothness of a solution is affected by the type of equation. The solutions to elliptic PDEs are smooth. In addition, boundary conditions at any point affect the solution at all points in the computational domain. Elliptic PDEs often describe the steady-state condition of a variable, for example a diffusion process that has reached equilibrium, or a steady-state temperature distribution. In hyperbolic PDEs, the smoothness of a solution depends on the initial and boundary conditions. For instance, if there is a jump in the data at the start or at the boundaries, this jump will propagate in the solution. Parabolic PDEs arise in time-dependent diffusion problems, for example the transient flow of heat.

6.3.2   Finite difference solution of parabolic equations

Equation (6.70) is an example of a parabolic partial differential equation:

$$\frac{\partial u}{\partial t} = D \frac{\partial^2 u}{\partial x^2}. \tag{6.70}$$

For instance, Equation (6.70) can describe the diffusion of a chemical species in porous material as a function of time and space, with diffusivity constant D. The equation can also describe the heat conduction in a material. As such it can be used to quantify how temperature is conducted throughout a material in time, with the heat diffusivity $D = \lambda/(\rho c_p)$. We limit the discussion to how to solve parabolic PDEs by using the finite difference method.

6.3.3   Forward difference method

The basic idea of the finite difference method for solving PDEs is to use a grid in the independent variables and to discretize the PDEs by introducing difference-quotient approximations, thereby reducing the problem to solving a system of equations. The first derivative in time can be approximated by the forward difference formula,

$$u_t(x_i, t_j) = \frac{u(x_i, t_{j+1}) - u(x_i, t_j)}{k} + O(k), \tag{6.71}$$

and the second derivative to the x variable can be approximated by the central finite difference formula,

$$u_{xx}(x_i, t_j) = \frac{u(x_{i+1}, t_j) - 2u(x_i, t_j) + u(x_{i-1}, t_j)}{h^2} + O(h^2). \tag{6.72}$$

Substituting Equations (6.71) and (6.72) into the heat equation, Equation (6.70), yields

$$\frac{1}{k}(u_{i,j+1} - u_{i,j}) = \frac{D}{h^2}(u_{i+1,j} - 2u_{i,j} + u_{i-1,j}), \tag{6.73}$$

where $u_{i,j}$ is the numerical approximation to $u(x_i, t_j)$. This forward time central space (FTCS) discretization has the local error $O(k) + O(h^2)$, i.e. it is first-order accurate in

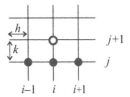

Figure 6.19. Mesh points for the forward difference method.

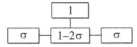

Figure 6.20. Stencil for the explicit forward method.

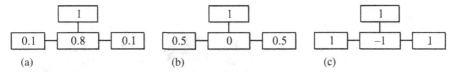

Figure 6.21. Stencil for the explicit forward method. (a) $\sigma = 0.1$; (b) $\sigma = 0.5$; (c) $\sigma = 1$.

time and second-order accurate in space. On rearranging, we obtain

$$u_{i,j+1} = u_{i,j} + \frac{kD}{h^2}(u_{i+1,j} - 2u_{i,j} + u_{i-1,j}).$$ (6.74)

Let us now introduce the parameter $\sigma = kD/h^2$; then Equation (6.74) can be written as

$$u_{i,j+1} = \sigma u_{i-1,j} + (1 - 2\sigma)u_{i,j} + \sigma u_{i+1,j}.$$ (6.75)

The forward difference formula is explicit as the new values can be determined directly from the previous known values, i.e. the solution takes the form of a marching procedure in time. The mesh points involved in stepping forward in time are shown in Figure 6.19. In the figure, the filled circles represent values known from the previous time step and the open circle is the unknown value.

In order to solve the PDE, the initial and boundary conditions also need to be specified. The explicit forward difference method is of little practical use because the method is only conditionally stable. This can be understood better by looking at the relationship between the unknown value and the values known from previous time steps, as in the stencil in Figure 6.20.

Note that when $\sigma = 0.5$ the solution at the new point is independent of the closest point (Figure 6.21(b)). When $\sigma > 0.5$, the new point depends negatively on the closest point (Figure 6.21(c)). A consequence of this is that the solution is unstable for $\sigma > 0.5$.

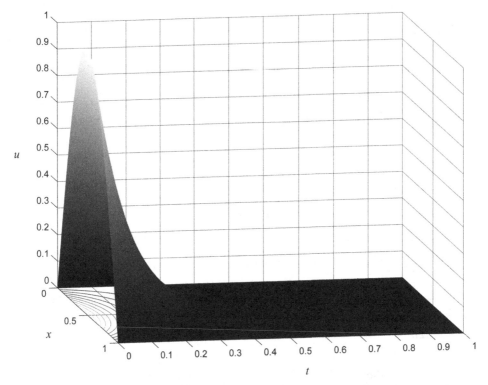

Figure 6.22. Solution of the heat equation problem, $D = 1.0$ and $\sigma < 0.5$, Example 6.9.

Example 6.9 Finite difference method – stability constraints

Consider solving the following problem:

$$
\begin{cases}
\dfrac{\partial u}{\partial t} = D\dfrac{\partial^2 u}{\partial x^2}, \\
x \in [0, 1], t \geq 0,
\end{cases}
\tag{6.76}
$$

with the simple Dirichlet boundary conditions

$$
\begin{aligned}
u(0, t) &= 0, \\
u(1, t) &= 0,
\end{aligned}
\tag{6.77}
$$

and the initial condition

$$
u(x, 0) = \sin(\pi x).
\tag{6.78}
$$

The solution of the heat equation problem ($D = 1.0$), using the explicit forward method, is shown in Figure 6.22. Here, $\sigma < 0.5$ and the solution is stable. Considering that the time scale for diffusive heat transport is $\tau = L^2/D$, the temperature profile is expected.

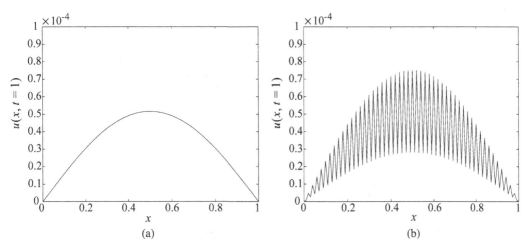

Figure 6.23. Stability of the forward explicit method for the heat equation. Temperature distributions are shown at $t = 1$; (a) stable for $\sigma = 0.4$; (b) unstable for $\sigma = 0.5005$.

The instability in the solution of the heat equation problem occurs for $\sigma > 0.5$. Figure 6.23(a) shows the temperature profile at $t = 1.0$ for $D = 1.0$ and $\sigma = 0.4$, and Figure 6.23(b) shows the predicted profile for $\sigma = 0.5005$. The instability for $\sigma > 0.5$ is very clear. The problem with stability can be overcome by using other difference methods, e.g. implicit methods.

6.3.4 Backward difference method

The implicit backward difference formula,

$$u_t(x_i, t_j) = \frac{u(x_i, t_j) - u(x_i, t_{j-1})}{k} + O(k), \tag{6.79}$$

improves stability. Upon substitution into the heat equation and rearrangement, the PDE is approximated numerically by

$$\frac{1}{k}(u_{i,j} - u_{i,j-1}) = \frac{D}{h^2}(u_{i+1,j} - 2u_{i,j} + u_{i-1,j}). \tag{6.80}$$

Stability analysis shows that the backward implicit method is stable for any choice of step size k and h. This means that the method is unconditionally stable; as such, the stability does not depend on the $\sigma = kD/h^2$ value.

As the local error is the same as in the forward difference method, i.e. $O(k) + O(h^2)$, the error from the time discretization dominates, assuming step sizes of $h \approx k$. Although the method is unconditionally stable, its accuracy is low due to the large local truncation error. Overall, the backward difference method is $O(k)$, which means that the error at a certain point decreases linearly with k. In other words, if k is cut in half, so is the error. In contrast, if h is cut in half (for a given k), the error would be about the same, and the amount of computation would increase.

6.3.5   Crank–Nicolson method

The Crank–Nicolson method is a combination of implicit and explicit methods, and the benefit of the method is twofold. First, the method is unconditionally stable, and second the truncation error is smaller, $O(k^2) + O(h^2)$. Implementing the Crank–Nicolson method in the heat equation gives

$$2u_{i,j} - 2u_{i,j-1} = \frac{Dk}{h^2}(u_{i+1,j} - 2u_{i,j} + u_{i-1,j} + u_{i+1,j-1} - 2u_{i,j-1} + u_{i-1,j-1}).$$

(6.81)

This section has provided a short introduction to solving PDEs using the finite difference method. But there is much more to explore, and the reader is referred to books on advanced numerical methods to learn more about how to solve PDEs. In most cases the reader will use existing numerical algorithms for solving ODEs and PDEs, either commercial or open-source. Section 6.4 gives a short description of some of these software products.

6.4   Simulation software

There are many numerical software products available, both commerical and open-source. It is good practice to use the available algorithms for solving ODEs and PDEs rather than to develop new ones from scratch. In this section we list some of most popular algorithms in MATLAB, since it is used widely in both academia and industry. Another popular software product is GNU Octave (freeware). Besides these software products, there is a series of useful books on numerical analysis, "Numerical Recipes."

6.4.1   MATLAB

The two work horses used for solving ODE problems are ode45 for non-stiff equations and ode15s for stiff equations. These methods have in common that the step size does not need to be specified, because the solvers include an estimate of the error at each step. The solver then adapts the step size to meet the specified error tolerance. The explicit solvers are all unstable with stiff systems. The solver will shorten time steps and the solution will take a long time, or the step reduces to a point where machine precision causes the algorithm to fail. The most widely used ODE solvers are classified in accordance to their characteristics, as discussed in the preceding sections, and as shown in Table 6.5.

The MATLAB ODE solvers are given three arguments: the function evaluating the right-hand side of the differential equation; the time interval to integrate over; and the initial conditions. There are also other important characteristics, besides the ones listed in the table, that might be of importance when selecting a particular algorithm. For example, the ode15s solver can solve differential algebraic equations (DAEs).

Methods for solving BVPs and PDEs are summarized in Table 6.6. The standard MATLAB installation has one PDE solver. But there is also a toolbox for solving PDEs

Table 6.5. Characteristics of MATLAB ODE-IVP solvers

Solver	Explicit/ implicit	Single step/ multistep	Non-stiff/stiff	Accuracy	Method
ode23	explicit	single	non-stiff	low	Runge–Kutta pair of order 2 and 3, of Bogacki and Shampine
ode45	explicit	single	non-stiff	medium	Runge–Kutta pair of order 4 and 5, of Dormand–Prince
ode113	explicit	multi	non-stiff	variable low to high	Adams–Bashforth–Moulton predictor–corrector pairs of order 1 to 13
ode15s	implict	multi	stiff	variable low to medium	numerical differentiation formulas, orders 1 to 5
ode23s	implicit	single	stiff	low	modified Rosenbrock formula of order 2

Table 6.6. BVP and PDE solvers

Method	Category	Comment
bvp4c	boundary value	solves two-point BVPs for ODEs
pdepe	parabolic and elliptic, PDEs	PDE solver for parabolic-elliptic PDEs
pdetool	PDEs	graphical user interface to solve PDEs

Table 6.7. Minimization and root-finding methods

Method	Comment
fzero	finds zero of single-variable function
fsolve	solves system of equations of several variables
fminunc	finds a minimum of a scalar function of several variables, unconstrained
fmincon	finds a constrained minimum of a scalar function of several variables
lsqnonlin	solves non-linear least square problems, e.g. regression analysis (see Chapter 7)

in which the command pdetool invokes a graphical user interface (GUI). In this GUI, the user can draw geometry, define boundary conditions, specify the PDE, generate the mesh, and compute and display the solution. It allows the solution of PDEs in two space dimensions and time.

6.4.2   Miscellaneous MATLAB algorithms

For the convenience of MATLAB beginners, we list some other useful algorithms in the Tables 6.7 and 6.8. Some of these algorithms and commands might assist in completing the practice problems.

Linear equation systems can be solved with the linsolve algorithm. Various post-processing and visualization commands are available, as shown in Table 6.8.

Table 6.8. Visualization of the solution

Method	Comment
plot	linear plot, 2D
loglog	log-log plot, 2D
contour3	creates a 3D contour plot of a surface
surf	colored surface plot, 3D

Analytical solutions of some ODEs can be obtained with dsolve. As an example, a solution to a particle settling at low particle Reynolds number can be obtained with the "Symbolic toolbox" in MATLAB:

```
y=dsolve('Dy=k1-k2*y','y(0)=0','t')
```

which gives

```
y = (k1 - k1/exp(k2*t))/k2
```

6.4.3   An example of MATLAB code

The solution to the problem in Example 6.1,

$$
\begin{cases}
\dfrac{dy}{dx} = f(x, y) = y + x, \\
y(0) = 1, \\
x \in [0, 1],
\end{cases}
$$

can be obtained by the following commands in the MATLAB prompt.

```
opts = odeset('RelTol', 1e-4);
[x,y] = ode45(@f, [0 1], 1, opts)
plot(x,y)
```

and by specifying the differential equation in a function file (m-file),

```
function dy = f(x,y)
dy = y + x;
```

Note that the MATLAB ODE solver is given three arguments: the function evaluating the right-hand side of the differential equation, the time interval to integrate over, and the initial conditions. The first line of code (opts) sets the relative error tolerance to 1e−4 (default is 1e−3). This line is followed by a command which calls the ode45 solver, and specifies that the differential equation is specified in the m-file. Further, the command also specifies that the ODE should be solved in the interval 0–1 and at the initial condition $y_0 = 1$. The third line is just used to plot the solution in a graph.

It is also possible to solve the problem by defining the function $f(x, y)$ "inline":

```
[x,y] = ode45(inline('y+x','x','y'), [0 1], 1, opts)
```

6.4.4   GNU Octave

Octave is a programming language with syntax very similar to that in MATLAB. It is used in academia and industry for numerical analysis. The software is available for free under the terms of the GNU General Public License and runs on both Windows and Unix systems.

6.5   Summary

Differential equations play a dominant role in modeling chemical engineering systems. This chapter has explained the principles underlying solutions to differential equations. The purpose was to introduce the concepts of accuracy, stability, and computational efficiency. With this knowledge, the reader can understand how different algorithms work, and how to select appropriate numerical algorithms for different kinds of differential equations. Readers who are particularly interested in this area are encouraged to study more advanced books on numerical analysis for more details.

6.6   Questions

(1) What is the difference between the local truncation error and the global error, and how are they related?

(2) How can error tolerance be ensured when a true solution does not exist when solving differential equations?

(3) Explain what is meant by adaptive step size methods, and why they are used.

(4) What characterizes multistep methods, and why are they often used?

(5) How do predictor–corrector pairs work, and why are they used?

(6) How is stability related to explicit and implicit methods, and what is a conditionally stable method?

(7) Explain what is meant by stiff differential equations, and what category of ODE solvers is suitable for stiff systems.

(8) Why are different solution methods needed for IVPs and BVPs, and how can BVPs be solved?

(9) Explain the principle behind the finite difference method.

(10) What do the collocation and finite element methods have in common, and how do they differ?

6.7   Practice problems

There are many books on numerical methods available that contain exercises that allow you to practice writing your own algorithms to solve differential equations. It is equally important to learn how to use the existing software products for numerical analysis, and select appropriate numerical methods for stability and efficiency reasons. With this in mind, some of the problems in this section require that you work with software for numerical computing, e.g. MATLAB.

6.1 Find the approximate solution to the IVP $dy/dt = t - y$, $y(0) = 1$, for $t \in [0, 1]$, using step sizes $h = 0.1, 0.01$, and 0.001. Determine the error at $t = 1$ and make a log-log plot of the global error as a function of the step size h, using
  (a) the forward Euler method,
  (b) the fourth-order Runge–Kutta method.
  (c) Confirm that the two lines are consistent with the order of the method.
  The analytical solution is $y(t) = t + 2/e^t - 1$.

6.2 Apply the explicit and implicit Euler methods to find an approximate solution to the following IVPs:
  (a) $\dfrac{dy}{dt} = 50(1 - y)$, $y(0) = 0.5$, $t \in [0, 1]$;

  (b) $\dfrac{dy}{dt} = 50(y - y^2)$, $y(0) = 0.5$, $t \in [0, 1]$.
  Plot the solutions for different step sizes. What step size is required for the explicit method to converge to the equilibrium value? Recall that Newton's method to find a root of $z(x) = 0$ and starting guess, x_0, is $x_{i+1} = x_i - [z(x_i)/z'(x_i)]$.

6.3 Write the higher-order differential equation as a system of first-order differential equations:
  (a) $\dfrac{d^2y}{dx^2} + x^2 \dfrac{dy}{dx} - y\sin(x) = 0$;

  (b) $\dfrac{d^2y}{dx^2} - y\dfrac{dy}{dx} + x^2 - \alpha = 0$;

  (c) $\dfrac{d^3y}{dx^3} - \dfrac{d^2y}{dx^2}\dfrac{dy}{dx} + y^2 - e^x = 0$;

  (d) $\dfrac{d^4y}{dx^4} - \dfrac{dy}{dx}y^2 + e^x = 0$.

6.4 Depending on the rate constants, chemical reacting systems might be stiff. One well-known stiff system is the Robertson example, which consists of non-linear ODEs characterized by a large difference in the rate constants. The three reactions occurring in the system are

$$A \xrightarrow{k_1} B,$$
$$B + B \xrightarrow{k_2} B + C,$$
$$B + C \xrightarrow{k_3} A + C,$$

where the rate constants are $k_1 = 0.04$, $k_2 = 3 \cdot 10^7$, $k_3 = 1 \cdot 10^4$, and the initial concentrations of the three species are $C_{A0} = 1$, $C_{B0} = 0$, and $C_{C0} = 0$.

(a) Assume the reactions are carried out in a batch reactor operated at isothermal conditions, and derive the system of ODEs.

(b) Use a solver suitable for stiff problems and plot the solution for $t \in [0, 100]$.

(c) Evaluate the performance of non-stiff and stiff solvers, e.g. MATLAB ode45 and ode15s, by analyzing the number of steps required by the two different solvers.

(d) Determine how the order of the method for stiff solvers is related to the error tolerance specified.

6.5 An ideal batch reactor is used to hydrogenate compound A. The hydrogen is fed through a gas sparger and dissolves in the liquid phase, where it reacts with compound A,

$$A + H_2 \xrightarrow{r} B,$$

where the reaction rate is $r = kC_A C_{H_2}$ and $k = 10^{-3} \, \text{m}^3 \, \text{s}^{-1} \, \text{mol}^{-1}$.

The hydrogen mass transfer rate is dominated by the liquid film resistance. This allows the mass transfer rate to be modeled as $Na = k_L a(C_{H_2}^i - C_{H_2})$, where $k_L a = 10 \, \text{s}^{-1}$. The initial concentration of compound A is $C_A^0 = 10 [\text{mol m}^{-3}]$, and, as hydrogen is replenished to maintain the reactor pressure, $C_{H_2}^i = 0.1 [\text{mol m}^{-3}]$.

(a) Derive the material balances that describe how the concentrations C_A and C_{H_2} vary in time. Assume perfect mixing and isothermal conditions in the reactor.

(b) Estimate the time constants and select a suitable ODE solver.

(c) Make a plot that shows how the concentrations change in time.

6.6 Diffusion and reaction inside a spherical catalyst particle are described by the following BVP:

$$\frac{d^2 C}{dr^2} + \frac{2}{r} \frac{dC}{dr} - 4C = 0,$$

$$C(1) = 1,$$

$$\frac{dC}{dr}\bigg|_{r=0} = 0,$$

$$r \in [0, 1].$$

(a) Use the collocation method, with a collocation point at $r = 0.5$, to find an approximate solution. The result can be compared with the concentration profile shown in Figure 6.13.

(b) Use the MATLAB built-in BVP solver to find the solution to the problem.

6.7 Heat transport by conduction is described by the parabolic PDE

$$\frac{\partial T}{\partial t} = D \frac{\partial^2 T}{\partial x^2},$$

$$x \in [0, 1], \quad t \in [0, 1],$$

the boundary conditions $T(0, t) = 0$ and $T(1, t) = 0$, the initial condition $T(x, 0) = \sin^2(4\pi x)$, and the heat diffusivity $D = \lambda/\rho c_p = 1$. Find the solution by

(a) using the finite difference, forward time central space, method;
(b) using the finite difference, Crank–Nicolson, method;
(c) using the MATLAB built-in PDE solver "pdepe".

6.8 A metal rod initially at ambient air temperature, 25 °C, is connected on one side
to a wall at constant temperature 100 °C, and the other side is surrounded by air.
Heat conduction through the rod and losses by convection to the surroundings are
balanced by accumulation of thermal energy in the rod. The energy balance is given
by

$$\frac{\partial T}{\partial t} = \frac{\lambda}{\rho c_p} \frac{\partial^2 T}{\partial x^2} - k_1(T - T_s),$$

where $T_s = 25$ °C is the surrounding air temperature (constant) and $k_1 =$
$5.25 \cdot 10^{-7}\,\mathrm{s}^{-1}$ (depends amongst other things on the heat conduction in air and
the radius of the rod). The rod is 1 m long and made of silver; the corresponding
material data for silver are $\lambda = 424\,\mathrm{W\,m^{-1}\,K^{-1}}$, $\rho = 10\,500\,\mathrm{kg\,m^{-3}}$, and $c_p =$
$236\,\mathrm{J\,kg^{-1}\,K^{-1}}$.
(a) Propose a suitable boundary condition for the tip of the rod, which is in contact
with air.
(b) Estimate the time constant of the system.
(c) Determine how the temperature profile evolves in time.

7 Statistical analysis of mathematical models

7.1 Introduction

In chemical engineering, mathematical modeling is crucial in order to design equipment, choose proper operating conditions, regulate processes, etc. It is almost always necessary to use experimental data for model development. Figure 7.1(a) shows a data set and linear fit for these data. From this result, is easy to see that this line describes these data. However, from the data set shown in Figure 7.1(b), this is not so clear. The solid line represents the linear fit for these data, which is derived from regression analysis. By simply observing the data, it can be seen that either of the dashed lines could be possible fits. These results clearly show that it is not possible to determine parameters for models only by which line looks a good fit, but that a detailed statistical analysis is needed.

In this chapter, we start by describing linear regression, which is a method for determining parameters in a model. The accuracy of the parameters can be estimated by confidence intervals and regions, which will be discussed in Section 7.5. Correlation between parameters is often a major problem for large mathematical models, and the determination of so-called correlation matrices will be described. In more complex chemical engineering models, non-linear regression is required, and this is also described in this chapter.

7.2 Linear regression

Regression analysis is a statistical method for determining parameters in models. The simplest form is a first-order straight-line model, which can be described by

$$y = \beta_0 + \beta_1 x + \varepsilon, \tag{7.1}$$

where

y is the dependent variable that can also be called the response variable;
x is the predictor of y, also known as the regressor or independent variable;
β_i are parameters;
ε is the stochastic part, and describes the random error.

In this chapter, the vectors and matrices will be written in bold italic style. The straight-line model can be divided into two parts: the deterministic part, $E(y)$, and

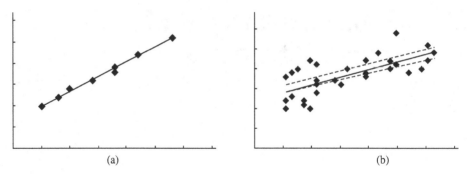

Figure 7.1. Data sets and corresponding linear fits.

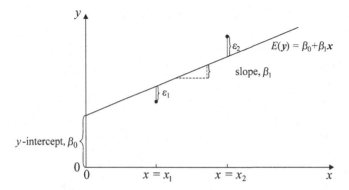

Figure 7.2. Simple linear regression model.

the stochastic part, ε. The term $E(y)$ is also called the expectation function, and it is described by

$$E(y) = \beta_0 + \beta_1 x. \tag{7.2}$$

Figure 7.2 shows a straight-line fit. The slope of the line represents β_1 and the y-intercept β_0 in Equation (7.1). The errors ε_1 for $x = x_1$ and ε_2 for $x = x_2$ are also shown in the figure.

A more general equation for the linear regression is the so-called multiple linear regression that is described by

$$y = \beta_0 + \beta_1 x_1 + \beta_2 x_2 + \cdots + \beta_k x_k + \varepsilon, \tag{7.3}$$

where x_1 and x_2 could be different variables for, e.g., pressure and temperature, but may also be powers of x, x^2, x^3, etc. This equation is denoted linear by virtue of its linear parameters. This can also be seen by the fact that $dy/d\beta$ is only a function of the independent variables, x. To illustrate linear models further, some examples are given. The model in Equation (7.4) is linear in its parameters and is called a polynomial model:

$$y = \beta_0 + \beta_1 x + \beta_2 x^2 + \beta_4 x^4 + \varepsilon. \tag{7.4}$$

Another set of models, which are also linear with respect to the parameters, are the so-called sinusoidal models:

$$y = \beta_0 + \beta_1 \sin(\theta) + \beta_2 \cos(\theta) + \varepsilon. \tag{7.5}$$

An example of a non-linear model is the Arrhenius equation, which is very important for describing the temperature dependence of the rate constants. It is given by

$$k = Ae^{-E_A/RT}, \tag{7.6}$$

where k is the rate constant, A is the pre-exponential factor, E_A is the activation barrier, R is the general gas constant, and T is the temperature. The rate constant is included in the material and heat balances for a system, with the result that the concentrations and temperature depend on the parameters in the rate constants. The parameters of this equation that should be determined are A and E_A. Rewriting Equation (7.6) using the notation in Equation (7.3) would result in

$$k = \beta_0 e^{-\beta_1/Rx}. \tag{7.7}$$

This yields

$$\frac{dk}{d\beta_1} = \frac{-\beta_0}{Rx} e^{-\beta_1/Rx},$$

and as this derivative contains β_1 it means that it is not linear in the parameter β_1; therefore this is a non-linear model.

7.2.1 Least squares method

The least squares method is a statistical approach used to estimate the parameters previously described. First, we will consider the simplest case, which is the first-order straight-line model. This model is described by

$$y = \beta_0 + \beta_1 x + \varepsilon. \tag{7.8}$$

The sum of square errors is denoted by SSE, or s, and is described by

$$SSE = \sum_{i=1}^{n} \varepsilon_i^2, \tag{7.9}$$

where ε_i is the random error at the experimental point i (see Figure 7.2). To illustrate this further, if there were experimental points $x_1, x_2, x_3, x_4, \ldots, x_n$, the corresponding errors would be described by $\varepsilon_1, \varepsilon_2, \varepsilon_3, \varepsilon_4, \ldots, \varepsilon_n$. These errors for experimental point i would be expressed by

$$\varepsilon_i = y_i - (\beta_0 + \beta_1 x_i). \tag{7.10}$$

This means that the error in each point is the difference between the experimentally measured value and the value calculated by the model. In the least square method, the SSE described in Equation (7.9) is minimized. The resulting line, when SSE is at a minimum, is known as the least square, or regression, line. The true model is described by

$$y = \beta_0 + \beta_1 x + \varepsilon, \tag{7.11}$$

and the estimated regression line can be described by

$$\hat{y} = b_0 + b_1 x, \tag{7.12}$$

where b_0 and b_1 are estimates of the true β_0 and β_1, respectively, and $\hat{y}$ is the y predicted by the model. When minimizing SSE, the normal equations are set up. This will be performed for the simplest case, the first-order straight-line model. The least square error is minimized when the derivative with respect to the two parameters equals zero:

$$\frac{\partial SSE}{\partial \beta_0} = 0, \tag{7.13}$$

$$\frac{\partial SSE}{\partial \beta_1} = 0, \tag{7.14}$$

where the least square error is given by

$$SSE = \sum_{i=1}^{n} \varepsilon_i^2 = \sum_{i=1}^{n} (y_i - (\beta_0 + \beta_1 x_i))^2. \tag{7.15}$$

When using the expression for SSE in Equation (7.15) and differentiating it with respect to each parameter, the following expressions are obtained:

$$\frac{\partial SSE}{\partial b_0} = \sum_{i=1}^{n} 2(y_i - (b_0 + b_1 x_i))(-1) = 0; \tag{7.16}$$

$$\frac{\partial SSE}{\partial b_1} = \sum_{i=1}^{n} 2(y_i - (b_0 + b_1 x_i))(-x_i)$$

$$= 2 \sum_{i=1}^{n} (y_i x_i - (b_0 x_i + b_1 x_i^2)) = 0. \tag{7.17}$$

Simplifications of Equations (7.16) and (7.17) result in

$$\sum_{i=1}^{n} (y_i) - n b_0 - b_1 \sum_{i=1}^{n} (x_i) = 0, \tag{7.18}$$

$$\sum_{i=1}^{n} (y_i x_i) - b_0 \sum_{i=1}^{n} (x_i) - b_1 \sum_{i=1}^{n} (x_i^2) = 0. \tag{7.19}$$

Equations (7.18) and (7.19) are often denoted as normal equations. The parameters b_0 and b_1 can be easily retrieved from these equations. Often, the averages of y and x, respectively, are used. They are denoted by $\bar{x}$ and $\bar{y}$, and are given by

$$\bar{x} = \frac{1}{n} \sum_{i=1}^{n} (x_i) \tag{7.20}$$

and

$$\bar{y} = \frac{1}{n}\sum_{i=1}^{n}(y_i). \tag{7.21}$$

This results in

$$b_0 = \bar{y} - b_1\bar{x} \tag{7.22}$$

and

$$b_1 = \frac{\sum_{i=1}^{n}(x_i - \bar{x})(y_i - \bar{y})}{\sum_{i=1}^{n}(x_i - \bar{x})^2}. \tag{7.23}$$

7.3 Linear regression in its generalized form

As discussed in Section 7.2, a multiple linear regression can be described by

$$y = \beta_0 + \beta_1 x_1 + \beta_2 x_2 + \cdots + \beta_k x_k + \varepsilon. \tag{7.24}$$

This equation can be written in a more generalized matrix form as follows:

$$y = X\beta + \varepsilon, \tag{7.25}$$

where the vectors and matrices y, X, β, and ε are defined as

$$y = \begin{bmatrix} y_1 \\ y_2 \\ y_3 \\ \vdots \\ y_n \end{bmatrix}, \quad X = \begin{bmatrix} 1 & x_{11} & x_{12} & x_{13} & \cdots & x_{1k} \\ 1 & x_{21} & x_{22} & x_{23} & \cdots & x_{2k} \\ 1 & x_{31} & x_{32} & x_{33} & \cdots & x_{3k} \\ \vdots & \vdots & \vdots & \vdots & & \vdots \\ 1 & x_{n1} & x_{n2} & x_{n3} & \cdots & x_{nk} \end{bmatrix}, \quad \beta = \begin{bmatrix} \beta_0 \\ \beta_1 \\ \beta_2 \\ \vdots \\ \beta_k \end{bmatrix}, \quad \varepsilon = \begin{bmatrix} \varepsilon_1 \\ \varepsilon_2 \\ \varepsilon_3 \\ \vdots \\ \varepsilon_n \end{bmatrix}. \tag{7.26}$$

The dimensions of these matrices are $n \times 1$ for y, $n \times p$ for X, $p \times 1$ for β, and $n \times 1$ for ε, where p is the number of parameters and is equal to $k + 1$ and n is the number of experimental points. As before, E is the expectation operator, and the expectation function is denoted by

$$E(y) = X\beta. \tag{7.27}$$

The vector of expected values for the error is $\mathbf{0}$, i.e.

$$E(\varepsilon) = \mathbf{0}. \tag{7.28}$$

The variance of the error is denoted by σ^2, and can be written using the variance, V:

$$V(\varepsilon) = \sigma^2. \tag{7.29}$$

In many experiments, the expected error is not constant for all observation points. We can still make a least squares minimization, but we cannot make any statistical analysis unless we stabilize the variance (see Section 7.4). For a detailed statistical analysis, each observation should be weighted with its variance.

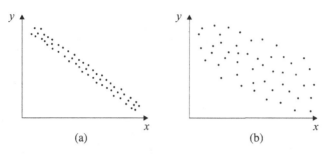

Figure 7.3. Two data sets with (a) small and (b) large variance, respectively.

To illustrate further the importance of the variance, two different data sets are shown in Figure 7.3. The data in Figure 7.3(a) have a small variance, whilst the data in Figure 7.3(b) have a large variance. The variance will be determined using so-called analysis of variance (ANOVA) tables, which will be described in detail in Section 7.8.2.

7.3.1   Least square method

The least square method is used for determining the parameters in the regression model, as described in Section 7.2.1. The sum of square error is given by

$$SSE = \sum_{i=1}^{n} \varepsilon_i^2. \tag{7.30}$$

When using models in a more generalized matrix form, the residual sum of squares is defined by

$$SSE = (y - X\beta)'(y - X\beta). \tag{7.31}$$

In the same way as for the simple case of the straight-line model, the sum of squares is minimized when the derivative of SSE with respect to the parameters is set at zero. This results in

$$\left.\frac{\partial SSE}{\partial \beta}\right|_b = 0, \tag{7.32}$$

where b ($p \times 1$) is a vector with the estimated parameters. In Equation (7.31), SSE can be further developed into

$$\begin{aligned} SSE &= y'y - y'X\beta - \beta'X'y + \beta'X'X\beta \\ &= y'y - 2\beta'X'y + \beta'X'X\beta. \end{aligned} \tag{7.33}$$

The derivative of SSE with respect to β gives

$$\left.\frac{\partial SSE}{\partial \beta}\right|_b = -2X'y + 2X'Xb = 0. \tag{7.34}$$

This is the normal equation in matrix form. This equation can be rearranged to yield

$$X'y = X'Xb, \tag{7.35}$$

which can be used to retrieve an equation for determining the parameter vector b:

$$b = (X'X)^{-1}X'y. \tag{7.36}$$

The basis for this equation is that the expected value for the error is 0, i.e.

$$E(\varepsilon) = 0, \tag{7.37}$$

and the variance of ε is σ^2, which can be written using the variance, V:

$$V(\varepsilon) = \sigma^2. \tag{7.38}$$

7.4 Weighted least squares

Weighted least squares are important and have several different applications. Examples of cases where weighted least squares are important include:

(1) stabilization of the variance;
(2) placing greater/less weight on certain experimental parts.

These two topics will be discussed in Sections 7.4.1 and 7.4.2.

7.4.1 Stabilization of the variance

Stabilizing the variance is the fundamental reason behind weighted residuals, where each experiment is weighted against its variance. The reason for doing this is that in many experiments the expected error is not equal in different observations. For these cases, in order to make a detailed statistical analysis we should weight each measurement with its variance. If adding weight factors, w_i, where the variance for each experiment is divided by the variance, this results in

$$w_i = \left(\frac{\sigma_i}{\sigma}\right)^2. \tag{7.39}$$

In many types of analytical equipment, the error is a fixed percentage of the maximum instrument reading, which results in a variance proportional to y_i, i.e. $\sigma_i \propto y_i$. This gives an uneven variance distribution that calls for using a weight factor, for example $1/y_i$, or a logarithm:

$$z_i = \ln y_i. \tag{7.40}$$

If z_i is differentiated, we obtain

$$dz_i = \frac{1}{y_i}dy_i, \tag{7.41}$$

which means that dz_i corresponds to a small error in y_i divided by y_i that yields the weight factor $1/y_i^2$. Weight factors are discussed further in the context of residual plots in Section 7.8.1.

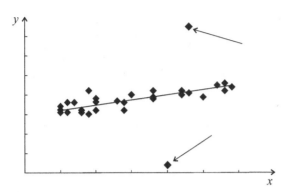

Figure 7.4. Two experimental outliers in a data set.

7.4.2   Placing greater/less weight on certain experimental parts

Another reason for using weight factors is to vary the weight according to different parts of an experiment. Figure 7.4 shows an example where a large set of data shows a clear linear trend. However, two experimental points deviate significantly (marked by arrows in Figure 7.4). These types of data points that exhibit clear deviations from the others are denoted "outliers" and are occasionally present in experimental series. If experimental outliers are found, the best procedure would be to repeat the experiment for these two points a few additional times to verify the observations. However, it is not always possible to redo the experiments. One possibility, for these cases, could be to place a lower weight on the error for these measuring points. It is also possible to remove these outliers in the regression analysis. However, if outliers were removed during the regression, it would be critical to add them to the figures, clearly indicating that the weight factor for these outliers would be zero (or another appropriate value).

The outliers in the example of Figure 7.4 do not substantially influence the regression model because one outlier is below the line and another is above the line. The dashed line in Figure 7.4 (which is on top of the solid line) shows the resulting regression line if both points were removed in the least square analysis. The fact that the two lines are roughly in the same position means that these outliers have only minor effect on the regression. Another case is shown in Figure 7.5, where there is a single outlier. In the same way as in Figure 7.4, the dashed line shows the regression line when removing the outlier (using a weight factor of zero on that point). In this case, this outlier has a clear impact on the least square estimation. It should be emphasized that outliers cannot be removed without a clear motivation, and that, if weight factors were used, they would have to be thoroughly described.

Recall the equation for the linear models,

$$y = \beta_0 + \beta_1 x_1 + \beta_2 x_2 + \cdots + \beta_k x_k + \varepsilon, \tag{7.42}$$

where

$$SSE = \sum_{i=1}^{n} \varepsilon_i^2 = \sum_{i-1}^{n} (y_i - \hat{y}_i)^2. \tag{7.43}$$

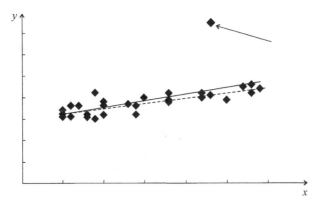

Figure 7.5. Effect of one experimental outlier on the regression model.

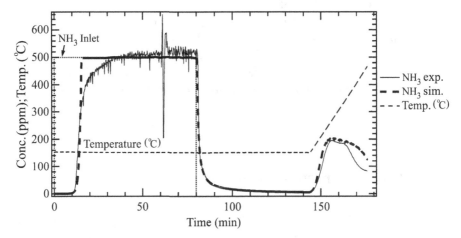

Figure 7.6. Model and experimental data for ammonia storage and desorption over a catalyst. From Olsson, L., Sjövall, H., and Blint, R. J., *Appl. Catal. B* **81**, 203–207, 2008. Reproduced with permission from Elsevier.

For the case using weight factors, the weighted sum of square errors, *WSSE*, can be expressed as follows:

$$WSSE = \sum_{i=1}^{n} w_i e_i^2 = \sum_{i=1}^{n} w_i (y_i - \hat{y}_i)^2. \tag{7.44}$$

In this section, we have described cases when experimental data points are not fully reliable, which might be the case with outliers. Another key area for weighted residuals is when certain parts of the experiment are very important and we would like to emphasize those parts of the regression model. An example from our research on automotive catalysis is depicted in Figure 7.6. In this experiment, the catalyst is exposed to ammonia for 80 min and the experimental (solid line) and simulated (dashed line) outlet ammonia concentration are shown in the figure. Initially, ammonia adsorbs on the catalyst, which is why the outlet ammonia concentration is zero. When the catalyst is saturated, ammonia breaks through. Thereafter, the ammonia is turned off and the temperature is increased, which results in an ammonia desorption peak. The binding strength can be received during the temperature ramp, so the desorption part is far more critical than the adsorption part when tuning a model to the data. However, the concentration in the desorption is quite low, with the result that, in a standard fitting procedure, the adsorption part influences the parameters in the model to a greater extent. Since the desorption part is far more important to the kinetic model, this is not the desired situation. In this case, weighted residuals can be used, where the error during the desorption part has a larger weight factor compared to the adsorption. In the example shown in Figure 7.6, the model is non-linear but the interpretation of weighted residuals remains the same.

Different weight factors can also be applied to the residual sum of squares when the response variables have varying orders of magnitude. For example, for models describing emission cleaning with the help of catalysis, it is common to use both molar fractions and temperatures as response variables. In many cases, the molar fraction is in the range 100×10^{-6}–500×10^{-6}, and the temperature is 400–700 K. If no weight factor were applied in this case, the errors in the prediction of the temperature would totally dominate and the molar fraction would have only a minor influence. However, as the errors in the molar fractions are crucial, this is not a good solution. Therefore, a weight factor can be applied to the molar fractions, with the result that the errors are similar in size to those for the temperatures (the opposite would also be possible using a smaller factor on the temperature instead).

7.5   Confidence intervals and regions

7.5.1   Confidence intervals

The basis for confidence intervals is that the error is normally distributed with the variance σ^2, i.e. $\boldsymbol{\varepsilon} \sim N(\mathbf{0}, \boldsymbol{I}\sigma^2)$. This also results in $\boldsymbol{b}$ being normally distributed and

$$E(\boldsymbol{b}) = \beta. \tag{7.45}$$

The variance, s^2, can be estimated by dividing the sum of square errors, SSE, by the degrees of freedom, v

$$v = n - p, \tag{7.46}$$

$$s^2 = \frac{SSE}{v} = \frac{SSE}{n - p} \tag{7.47}$$

where n is the number of experimental data points and p is the number of parameters in the model. Equation (7.46) applies to the case for which there is only one source of errors. The estimation of the degrees of freedom, v, will be discussed in detail in Section 7.5.1.1. The standard error of the parameter p is calculated according to:

$$se(b_p) = s\sqrt{\{(\boldsymbol{X}'\boldsymbol{X})^{-1}\}_{pp}}, \tag{7.48}$$

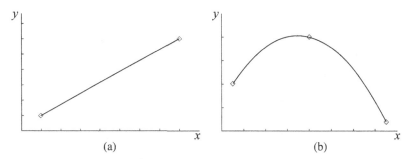

Figure 7.7. (a) Straight-line fit to two experimental points. (b) Quadratic fit to three experimental points.

where $\{(X'X)^{-1}\}_{pp}$ is the pth diagonal term of the matrix $(X'X)^{-1}$. For example, the standard error of parameter b_3 is

$$se(b_3) = s\sqrt{\{(X'X)^{-1}\}_{33}}, \tag{7.49}$$

where $\{(X'X)^{-1}\}_{33}$ is the third diagonal term in the matrix. By using these results, it is possible to determine the $1 - \alpha$ confidence interval for parameter b_p in the model as follows:

$$b_p \pm se(b_p)t(n - p; \alpha/2), \tag{7.50}$$

where $t(n - p; \alpha/2)$ is the Student's t-distribution with $n - p$ degrees of freedom. This distribution is described in Section 7.5.1.2. The confidence intervals described in Equation (7.50) represent one of the most common methods to determine confidence intervals. It is also possible to determine the $1 - \alpha$ confidence interval for the dependent variable $\hat{y}(x_0)$, which can also be denoted the "mean response" and written as $\mu(x_0)$:

$$x_0'b \pm s\sqrt{x_0'(X'X)^{-1}x_0}t(n - p; \alpha/2). \tag{7.51}$$

7.5.1.1 Number of degrees of freedom

The number of degrees of freedom is the number of observations of the random variation. When there is only one source of errors, e.g. the chemical analysis, the degrees of freedom are

$$v = n - p. \tag{7.52}$$

If there were other sources of the random variation that were more important than the spreading in repeated analyses, the number of observations of the dominant variation would be the number of degrees of freedom. When determining the confidence intervals, one step would be to calculate the number of degrees of freedom ($v = n - p$) as described. The reason why the number of parameters must be extracted from the number of degrees of freedom can be understood from the following example. In Figure 7.7(a), the result of a straight line fit ($\hat{y} = b_0 + b_1x$) to two points is shown. The fit is naturally

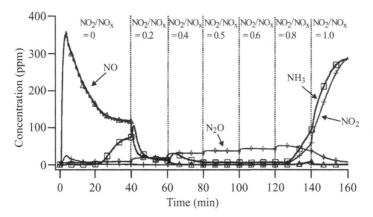

Figure 7.8. NO, NO$_2$, NH$_3$, and N$_2$O concentration after an ammonia selective catalytic reduction (SCR) experiment over a copper–zeolite catalyst. The inlet gas composition is 400 ppm NO$_x$ (NO or NO$_2$), 400 ppm NH$_3$, and 8% O$_2$. From Sjövall, H., Olsson, L., Fridell, E., and Blint, R. J., *Appl. Catal. B* **64**, 180–188, 2006. Reproduced with permission from Elsevier.

perfect because there is only one way in which a line can be drawn between two points. The number of degrees of freedom for this case is therefore

$$v = n - p = 2 - 2 = 0. \tag{7.53}$$

Thus there are no observations of the random error and consequently no degrees of freedom. Similar results can be obtained if a quadratic model $\left(\hat{y} = b_0 + b_1 x + b_2 x^2\right)$ were used to fit three experimental points. For this case, the number of degrees of freedom would be zero as well ($v = n - p = 3 - 3 = 0$) (see Figure 7.7(b)).

The number of parameters in the model, p, is often easily calculated. The degrees of freedom are calculated using $n - p$ (see Equation (7.53)), and this example illustrates the difficulties that can arise when we determine n and thereby v. However, the number of data points, n, is not always trivial to determine.

An example will be used to illustrate the number of degrees of freedom. In Figure 7.8, the experimental results from one ammonia selective catalytic reduction (SCR) experiment are shown. In this experiment, a copper–zeolite catalyst was exposed to ammonia, NO$_x$ (NO + NO$_2$) and oxygen, while varying the NO$_2$ to NO$_x$ ratio. The outlet concentrations of NO, NO$_2$, NH$_3$, and N$_2$O were measured and are depicted in Figure 7.8. If these data points were used in a model, what would the number of degrees of freedom be?

The experiment was 160 min long and measurement points were taken approximately every 6 s, resulting in 1600 data points for each of the four gases, i.e. a total of 6400 data points. However, this does not mean that $n = 6400$. This is because we have seven different conditions (varying NO$_2$ to NO$_x$ ratio) and each condition takes a long time to reach steady-state conditions. Thus, the concentrations are constant in the final part of the experiment for each condition and therefore n is significantly lower than 6400. However, since transient effects occur when the gas composition changes due to the

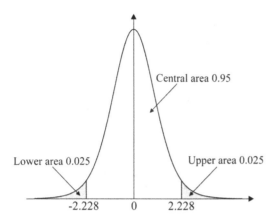

Figure 7.9. Student's t-distribution for 10 degrees of freedom, showing 95% confidence limit.

accumulation of species on the surfaces, n is much larger than $7 \times 4 = 28$ (seven conditions and four gases). In order to determine n, only the points where the condition is changed or when transient effects occur should be considered.

7.5.1.2   Student's t-distribution

As mentioned earlier, the t-distribution is used to determine confidence intervals. The t-distribution has two parameters: the number of degrees of freedom, v, and the significance level of the test divided by two ($\alpha/2$). We have already discussed v thoroughly. Figure 7.9 shows the t-distribution for a case with $v = 10$ degrees of freedom and $\alpha = 0.05$. The two tails represent an area of 0.025 each, which is $\alpha/2$ for each area. This results in a central area of $1 - 2 \times 0.025 = 0.95$. Thus, the confidence interval when using this t-value is at the 95% level. The values for the t-distribution can be found in Appendix C.

The t-distribution for 2, 10, and 1000 degrees of freedom is shown in Figure 7.10. The x-value for the 95% confidence limit for the three cases is marked. When increasing the degrees of freedom, and the number of observations of the random variation is increasing, the t-value becomes lower. This results in lower confidence intervals. Note that when the degrees of freedom $\rightarrow \infty$, the t-distribution will equal the normal distribution.

7.5.2   Student's t-tests of individual parameters

It is possible also to conduct t-tests on individual parameters. For this purpose, t_{obs} is calculated according to

$$t_{obs} = \frac{b_p}{se(b_p)}, \tag{7.54}$$

and this value is compared with the t-value from the t-distribution with $n - p$ degrees of freedom, $t(n - p; \alpha/2)$. If $t_{obs} > t(n - p; \alpha/2)$, the parameter is significant.

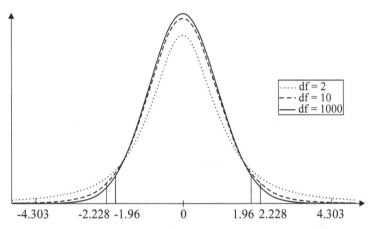

Figure 7.10. Student's t-distribution for 2, 10, or 1000 degrees of freedom, showing 95% confidence limit.

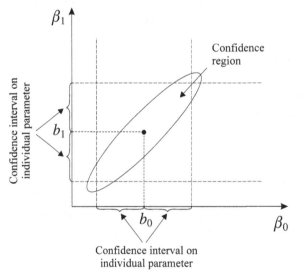

Figure 7.11. Individual confidence intervals and joint confidence region for a two-parameter model ($\hat{y} = b_0 + b_1 x$).

7.5.3   Confidence regions and bands

Confidence intervals were described in Section 7.5.1. These intervals are individual for each parameter. The individual confidence intervals for a two-parameter model ($\hat{y} = b_0 + b_1 x$) are shown in Figure 7.11, where the dashed lines show the confidence interval for each parameter. It is easy to assume that the area inside the dashed rectangle is the confidence region for the combination of these parameters; however, this is not the case. The joint confidence region for the two parameters is also depicted in the figure. The confidence region accounts for the variation of both parameters simultaneously,

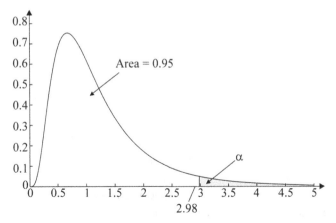

Figure 7.12. F-distribution for 10 and 10 degrees of freedom, showing 95% confidence limit.

which is indicated by the solid line in Figure 7.11. The joint confidence region is ellipsoidal, whereas the individual levels form a rectangle, differences that are important to notice.

To calculate the $1 - \alpha$ confidence region, the following relationship can be used for the linear case:

$$(\boldsymbol{\beta} - \boldsymbol{b})'X'X(\boldsymbol{\beta} - \boldsymbol{b}) \le ps^2 F(p, n - p; \alpha). \tag{7.55}$$

This yields an elliptical contour. In more general terms, the $1 - \alpha$ confidence region is determined by

$$SSE(\boldsymbol{\beta}) \le SSE(\boldsymbol{b}) \left[1 + \frac{p}{n - p} F(p, n - p; \alpha) \right], \tag{7.56}$$

where $F(p, n - p; \alpha)$ is the F-distribution (see Appendix C) for the degrees of freedom, p, and $n - p$, respectively, at level α. For linear models, the results for the confidence regions will be the same when using Equation (7.55) or Equation (7.56). Note that for the F-distribution the level to be used is α, whilst for the t-distribution it is $\alpha/2$. The reason for this difference is that the t-distribution is centered around zero and has two tails, whereas the F-distribution is non-symmetrical (and positive), and only has one tail (Figure 7.12).

As described in Section 7.5.1, the $1 - \alpha$ confidence interval for the dependent variable $\hat{y}(x_0)$ is determined according to

$$x_0'b \pm s\sqrt{x_0'(X'X)^{-1}x_0} \; t(n - p; \alpha/2). \tag{7.57}$$

If this calculation is performed for several different points (x), a "point-wise confidence band" is retrieved. It is also possible to determine the $1 - \alpha$ confidence interval for the value predicted, which is obtained when considering all intervals simultaneously, i.e.

$$x'b \pm s\sqrt{x'(X'X)^{-1}x} \sqrt{pF(p; n - p; \alpha)}. \tag{7.58}$$

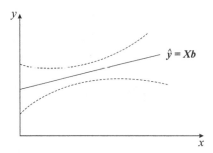

Figure 7.13. Schematic description of the confidence band.

This is denoted the "simultaneous $1 - \alpha$ confidence band" for the response function at any x. Note that the confidence band (Figure 7.13) depends on the independent variable x.

7.6   Correlation between parameters

After a model has been constructed, it is important to determine the correlation between the parameters. If there is high correlation and one parameter is changed, another parameter must also change. Multiple choices of parameters can result in a similar sum of square errors. In order to determine the correlation between parameters, the variance and co-variance must be calculated. Therefore, we start with a description of the variance calculation, followed by the determination of the co-variance, and finally the correlation matrix.

7.6.1   Variance and co-variance

The variance matrix for the parameters consists of the variance for each parameter in the diagonal and co-variances in the other positions. Equation (7.59) contains an example of a model with four parameters (b_0, b_1, b_2, and b_3) shown to illustrate this relationship:

$$V(b) = V \begin{pmatrix} b_0 \\ b_1 \\ b_2 \\ b_3 \end{pmatrix}$$

$$= \begin{bmatrix} V(b_0) & cov(b_0, b_1) & cov(b_0, b_2) & cov(b_0, b_3) \\ cov(b_1, b_0) & V(b_1) & cov(b_1, b_2) & cov(b_1, b_3) \\ cov(b_2, b_0) & cov(b_2, b_1) & V(b_2) & cov(b_2, b_3) \\ cov(b_3, b_0) & cov(b_3, b_1) & cov(b_3, b_2) & V(b_3) \end{bmatrix}. \qquad (7.59)$$

The variance matrix can be determined by the following correlation:

$$V(b) = (X'X)^{-1}\sigma^2. \qquad (7.60)$$

For the cases when σ^2 is unknown, the estimated variance, s^2, can be used instead.

7.6.2 Correlation matrix

The correlation matrix, C, can be retrieved from the variance matrix according to the following equation:

$$C_{ij} = \frac{cov(b_i, b_j)}{\sqrt{V(b_i)V(b_j)}}.$$ (7.61)

Combining Equations (7.60) and (7.61) results in the following expression:

$$C_{ij} = \frac{\{(X'X)^{-1}\}_{ij}}{\sqrt{\{(X'X)^{-1}\}_{ii}\{(X'X)^{-1}\}_{jj}}},$$ (7.62)

which can be used for determining the correlation.

Each component in C_{ij} is within the range -1 to 1. If the correlation equals 0, there would be no correlation between the parameters, and if the correlation is close to 1 or -1, the parameters are highly correlated. This can be a consequence of:

(1) Poor experimental design.
(2) Over-parameterizing, i.e. there are too many parameters and not all of them can be resolved. Over-parameterization can be estimated using the rank of the $X'X$-matrix (or the $J'J$-matrix, for the non-linear case; the Jacobian, J, will be described in Section 7.7.3).
(3) For non-linear models, there can be built-in correlations. An example is when the Arrhenius equation is used for determining the rate constant;

$$k = Ae^{-E_A/RT},$$ (7.63)

where k is the rate constant, A is the pre-exponential factor, E_A is the activation barrier, R is the general gas constant, and T is the temperature. The parameters that are used in the modeling are A and E_A, which are always correlated, at least to some extent.
(4) High correlation can also be obtained if the independent variables are not centered in a polynomial. For example, in the model

$$\hat{y} = b_0 + b_1 x + b_2 x^2 + b_3 x^3 + b_4 x^4,$$ (7.64)

the parameters b_0, b_1, b_2, b_3, and b_4 are correlated because x is always correlated with x^2, x^3, and x^4 if $x > 0$, for all x. This model can be centered by

$$x_{centered} = x - \bar{x},$$ (7.65)

where $\bar{x}$ is the mean of x. However, even for this model, there is a correlation between b_0 and b_2, because $(x - \bar{x})^2 > 0$ for all x. This is also the case for x^4. Likewise, in the non-linear case, it is recommended that the data are centered in order to decrease the correlation between the parameters. One example is the Arrhenius equation, Equation (7.63), in which the temperature can be centered by using a reference temperature, T_{ref}, which can be chosen to be the mean temperature. This gives

$$k = A \exp \left[\frac{E_A}{R} \left(\frac{1}{T} - \frac{1}{T_{ref}} \right) \right]$$ (7.66)

and results in a lower correlation between the parameters when doing the regression analysis. This transformation will not affect the activation energy, but the pre-exponential factor will be the rate constant at $T = T_{ref}$.

7.7   Non-linear regression

In chemical engineering, many models are non-linear, which means that they are non-linear with respect to the parameters. This will be illustrated by a few examples below. The simple straight-line model,

$$\hat{y} = b_0 + b_1 x, \tag{7.67}$$

is of course linear. However, the following models (because they are linear in the parameters) are also linear. Note that the regressor, x, can have different exponents, sine form, etc., while the model remains linear. We have

$$y = \beta_0 + \beta_1 x + \beta_2 x^2 + \beta_3 x^4 + \beta x^6; \tag{7.68}$$

$$y = \beta_0 + \beta_1 \sin(x) + \beta_2 \cos(x). \tag{7.69}$$

However, due to the form of the parameters, the following models are non-linear:

$$y = \theta_0 + e^{\theta_1 x}; \tag{7.70}$$

$$y = \theta_0 + \theta_1 \sin(\theta_2 x) + \theta_3 \cos(\theta_4 x). \tag{7.71}$$

In order to separate clearly the linear and non-linear models, we denote the parameters for the linear models as β_i, and that for the non-linear models as θ_i. In this chapter, we discuss the statistical evaluation of non-linear models, which is often very complex; therefore, iterative numerical solutions will be used.

7.7.1   Intrinsically linear models

The models described in Equations (7.70) and (7.71) are non-linear. It is possible to transform some non-linear models into linear models; these types of models are referred to as "intrinsically linear." It is important to note that the disturbance term is also transformed when transforming a model. We present two examples of models that can be transformed and are thereby intrinsically linear. For catalytic reactions, the reaction rate may contain an inhibition term in the denominator. The reaction rate at constant temperature can, for some cases, be expressed as

$$f(x, \theta) = \frac{\theta_1 x}{1 + \theta_2 x}. \tag{7.72}$$

This equation can be transformed into

$$y = \frac{1}{f(x, \theta)} = \frac{1 + \theta_2 x}{\theta_1 x} = \frac{\theta_2}{\theta_1} + \frac{1}{\theta_1} \frac{1}{x} = \gamma_1 + \gamma_2 u. \tag{7.73}$$

Another example is

$$f(x, \theta) = \exp(\theta_1 + \theta_2 x + \theta_3 x^2). \tag{7.74}$$

This equation can be transformed into

$$y = \ln(f(x, \theta)) = \theta_1 + \theta_2 x + \theta_3 x^2 = \gamma_1 + \gamma_2 x + \gamma_3 x^2. \tag{7.75}$$

When using logarithms or, more generally, other transformations, it is important to note that the weights for the different measuring points will differ. The two examples of models given here can be transformed and are therefore intrinsically linear. However, non-linear models cannot in most cases be transformed; these are discussed in Section 7.7.2.

7.7.2 Non-linear models

Non-linear models can be expressed in matrix form as

$$y = f(x, \theta) + \varepsilon. \tag{7.76}$$

The error sum of squares for the non-linear case can be expressed as

$$SSE(\theta) = \sum_{i=1}^{n} \varepsilon_i^2 = \sum_{i=1}^{n} (y_i - f(x_i, \hat{\theta}))^2. \tag{7.77}$$

In the same way as for the linear case, the normal equations are set up by differentiating $SSE(\theta)$ with respect to the parameters and setting the equations equal to zero. This will result in an estimate of θ, which we denote $\hat{\theta}$; i.e.

$$\frac{\partial SSE(\theta)}{\partial \theta} = 2 \sum_{i=1}^{n} \{y_i - f(x_i, \hat{\theta})\} \left[\frac{\partial f(x_i, \hat{\theta})}{\partial \theta_p} \right]_{\theta = \hat{\theta}} = 0. \tag{7.78}$$

This is performed for each parameter, and results in p (the number of parameters) normal equations. These equations are solved by different numerical methods. In much of the available software, it is possible to set up $SSE(\theta)$ and then call for a routine that minimizes the error sum of squares. For example, this can be performed in MATLAB using the function lsqnonlin.

An early method used for solving these equations is the Gauss–Newton method, for which the Taylor series is used. Other methods include "steepest descent" and "Marquardt's compromise." It is not the scope of this book to go into the details of solving non-linear normal equations because they are often solved computationally by using various software. However, this information is readily available in various textbooks. In this book, we will use linearization to determine approximate confidence levels and correlation matrices. For many applications, linearization is a very important tool that is used as a standard method in research.

7.7.3   Approximate confidence levels and regions for non-linear models

An important method for calculating approximate confidence intervals is through linearization with a first-order Taylor series expansion around the estimated parameter $\hat{\theta}$, which results in

$$f(x_n, \theta) \approx f(x_n, \hat{\theta}) + \left[\frac{\partial f(x_n, \hat{\theta})}{\partial \theta_p} \right]_{\theta = \hat{\theta}} (\theta - \hat{\theta}). \qquad (7.79)$$

This can be compared to the linear case, where

$$f(x, \theta) = b_0 + b_1 \theta. \qquad (7.80)$$

The derivative

$$\left[\frac{\partial f(x_n, \hat{\theta})}{\partial \theta_p} \right]_{\theta = \hat{\theta}}$$

of Equation (7.79) is denoted Jacobian and can be written as

$$J_{n,p} = \left[\frac{\partial f(x_n, \hat{\theta})}{\partial \theta_p} \right]_{\theta = \hat{\theta}} \qquad (7.81)$$

where n is the experimental point and p is the parameter.

Practically, an approximation of the Jacobian can be determined by the following procedure. First, find the parameters ($\hat{\theta}$) that give a minimum $SSE(\theta)$. Change one parameter at a time by a certain amount, for example 1% of the parameter value, with this parameter denoted $\hat{\theta}_{p,1\%}$. Note that in $\hat{\theta}_{p,1\%}$, only one parameter, p, changes, while all others remain at the minima. Thereafter, change this parameter back, change the next one, and so on. The approximated Jacobian can then be retrieved by

$$J_{n,p} \approx \frac{f(x_n, \hat{\theta}) - f(x_n, \hat{\theta}_{p,1\%})}{\Delta \theta_p}. \qquad (7.82)$$

The $1 - \alpha$ approximate confidence interval for the parameter $\hat{\theta}_p$ in the model can be determined by

$$\hat{\theta}_p \pm se(\hat{\theta}_p) t(n - p; \alpha/2), \qquad (7.83)$$

where $t(n - p; \alpha/2)$ is the Student's t-distribution with $n - p$ degrees of freedom. The standard error of the parameter p, $se(\hat{\theta}_p)$, is calculated according to

$$se(\hat{\theta}_p) = s \sqrt{\{(J'J)^{-1}\}_{pp}}. \qquad (7.84)$$

Note the similarities between this equation and the case of the linear equation. The only difference is that X is replaced by J. The standard deviation, s, can be retrieved from the variance, s^2:

$$s^2 = \frac{SSE}{v} = \frac{SSE}{n - p}. \qquad (7.85)$$

In order to determine approximate confidence intervals, it is critical that the model is really at a minimum. If this is not the case, the model is offset to the minimum region with large derivatives, and when changing the parameter by 1% there would be a much larger change to the predicted model than if you were at a flat minimum. As a result, the Jacobian will be overestimated and the confidence intervals thereby underestimated. Thus, if you are not at the minimum and the above equations were used for determining the confidence intervals, severe errors may occur.

7.7.3.1 Confidence bands

As for the linear case, the $1 - \alpha$ confidence interval for the dependent variable $f(x_0, \hat{\theta})$, which can also be denoted the "mean response," can be determined by

$$f(x_0, \hat{\theta}) \pm s \sqrt{j_0'(J'J)^{-1}j_0} t(n - p; \alpha/2), \tag{7.86}$$

where

$$j_0 = \left[\frac{\partial f(x_0, \theta)}{\partial \theta} \right]_{\theta = \hat{\theta}}. \tag{7.87}$$

The confidence intervals, for several points, together form a so-called "$1 - \alpha$ pointwise confidence band." The "simultaneous confidence band" at the $1 - \alpha$ level can be determined according to

$$f(x, \hat{\theta}) \pm s \sqrt{j'(J'J)^{-1}j} \sqrt{p F(p; n - p; \alpha)}. \tag{7.88}$$

7.7.3.2 Confidence regions

The $1 - \alpha$ joined confidence region for a non-linear model can be determined according to

$$SSE(\theta) \leq SSE(\hat{\theta}) \left[1 + \frac{p}{n - p} F(p, n - p; \alpha) \right]. \tag{7.89}$$

This is the exact confidence region with approximate confidence levels. The confidence region considers the effect of varying all parameters. For linear models, the confidence regions assume an elliptical shape, whereas the region is often more banana shaped for non-linear models.

It is also possible to determine the approximate elliptical contour for non-linear models, with exact probability level $1 - \alpha$, using

$$(\theta - \hat{\theta})' \times J'J \times (\theta - \hat{\theta}) \leq p s^2 F(p, n - p, 1 - \alpha). \tag{7.90}$$

Note that this results in an elliptical confidence region and that this is only an approximation, not the exact region. The differences between these two confidence regions are illustrated in Case study 7.3 (see Section 7.11).

7.7.4   Correlation between parameters for non-linear models

The correlation matrix, C, can also be calculated for non-linear models by using the Jacobian J instead of the X-matrix. This results in the following equation:

$$C_{ij} = \frac{\{(J'J)^{-1}\}_{ij}}{\sqrt{\{(J'J)^{-1}\}_{ii}\{(J'J)^{-1}\}_{jj}}}. \tag{7.91}$$

In the same way as for the linear case, each component in the C-matrix is always in the -1 to 1 range. There is no correlation between the parameters if $C_{ij} = 0$. However, if the correlation is close to 1 or -1, the parameters are highly correlated.

7.8   Model assessments

After the model has been constructed and the parameters estimated, it is crucial to evaluate it. In this book, we describe two methods for evaluating models: residual plots and lack of fit. Residual plots are discussed in Section 7.8.1 and lack of fit is described in the context of the analysis of variance (ANOVA) table. The error is defined as

$$e = y - \hat{y}, \tag{7.92}$$

which refers to the differences between the experimentally measured values and the values predicted by the model. The assumptions usually made when developing a model through regression analysis are that (i) the error is random, (ii) the error is normally distributed, and (iii) the variance is constant and equal to σ^2. After the model has been constructed, it should be confirmed that none of these assumptions has been violated.

7.8.1   Residual plots

Different types of residual plots are constructed in order to check that there are no trends pertaining to the errors. The error can be plotted versus time, independent variable (X), predicted response variable ($\hat{y}$), in addition to other parameters, and are discussed below.

7.8.1.1   Overall residual plot

The residual should be normally distributed as described ($\varepsilon \sim N(0, \sigma)$). This means that if we divide by the variance, the following normal distribution would result:

$$\frac{\varepsilon}{\sigma} \sim N(0, 1), \tag{7.93}$$

where σ can be estimated by s (see Equation (7.85)), and thereby *standardized residuals* can be retrieved:

$$d = \frac{e}{s}. \tag{7.94}$$

Using the t-distribution, $t(n - p; \alpha/2)$, with an infinite number of degrees of freedom at the 95% level, ($\alpha = 0.05$, $\alpha/2 = 0.025$), the result would be 1.96, and at the 99% level ($\alpha = 0.001$, $\alpha/2 = 0.005$) the result would be 2.576. This means that approximately

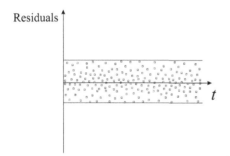

Figure 7.14. Residual plot for an example with randomly distributed errors.

95% of the standardized errors should be in the -1.96 to 1.96 range and approximately 99% in the -2.576 to 2.576 range. For other degrees of freedom, the t-values should be taken from the t-distribution (see Appendix C). However, because the variance is not constant for each error, this is not exact. Therefore, it is better to use the *Studentized residuals*,

$$z_i^* = \frac{e_i}{s\sqrt{1 - h_{ii}}}, \tag{7.95}$$

where h_{ii} is the "leverage" and denotes the diagonal elements of the Hat matrix, H:

$$H = X(X'X)^{-1}X'. \tag{7.96}$$

It is useful to construct residual plots based on the error, e, or the standardized residuals, for many applications.

7.8.1.2 Residual plot versus time

An important residual plot is when the residual is plotted against time. Figure 7.14 illustrates an example of errors plotted versus time and where the residuals show no trend. This is therefore the desired result in a residual plot.

Examples of other results for which the residuals are not random are shown in Figure 7.15. The results in Figure 7.15(a) show a clear linear trend; thus, with respect to time, a linear term is missing in the model. In Figure 7.15(b), a quadratic shape is observed; thus, a term containing t^2 must be added to the model. Another type of deviation is shown in Figure 7.15(c), where the error is increasing over time. The reason for this is an increasing variance, which can be compensated for using weighted residuals. Also, a transformation might be a solution. Finally, the residual plot in Figure 7.15(d) illustrates a rather complex behavior with increasing variance which occasionally drops down to low values, with the performance repeated. With this type of pattern, it is important to check what happened experimentally at the time of the sudden drop in variance. An example where this behavior might happen is when components are measured by means of an analyzer that has a drift in error over time. This means that the error becomes increasingly larger from the time when the calibration of the instrument was conducted. When a new calibration is made, the error drops down again. What should be done in this case is to calibrate the instrument more often, in order not to experience

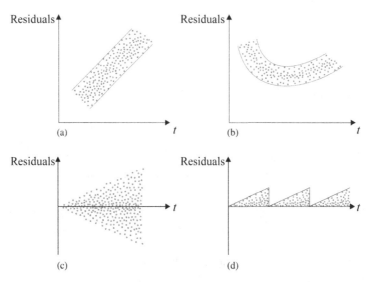

Figure 7.15. Residual plot versus time for examples with non-random errors.

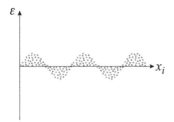

Figure 7.16. Residual plot versus the regressor variable (x_i), where the errors are not random.

this problem. This example clearly shows that all trends observed in the residual plots should not be removed by adding new terms to the model. For some cases, there are experimental problems, resulting in experiments having to be repeated or the experimental procedure refined, representing the causes of the trends in residual plots. It is therefore crucial for the person conducting the experiment to collaborate closely with the model developer. There are many additional types of patterns that the residual plots can assume; the objective of Figure 7.15 was to illustrate some of these patterns.

7.8.1.3   Residual plot versus x_i

The residual can also be plotted versus each regressor variable (x_i). These plots reveal any missing trends in the model with respect to that regressor variable. The results shown in Figure 7.15(a) and (b) could also occur when plotting the residual versus x_i, which indicates that a linear or quadratic term in x_i is missing. Another example is shown in Figure 7.16, where a sinusoidal term could be added.

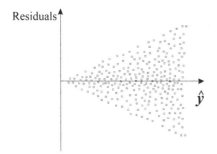

Figure 7.17. Residual plot versus $\hat{y}$ calculated from the model, where the errors are not random.

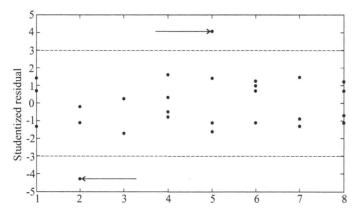

Figure 7.18. Studentized residuals.

7.8.1.4 Residual plot versus $\hat{y}$

The residual can also be plotted versus $\hat{y}$ calculated from the model. This type of residual plot is shown in Figure 7.17. The results show that the variance is increasing when the predicted value increases. A reason for this may be that a lot of experimental analysis equipment has errors that are a percentage level of the value, which means that the error increases when the value increases. Weighted residuals may represent a solution to such a case, where the weights decrease as the $\hat{y}$ value increases. A transformation of the model might be another alternative.

7.8.1.5 Outliers

Outliers are experimental points that deviate significantly from the others in the plot. This phenomenon was discussed in Section 7.4.2, and examples are shown in Figure 7.4 and Figure 7.5. If the Studentized residuals z_i^* are plotted, they should be in the -3 to 3 range. However, if there are points larger than 3 or smaller than -3, they represent outliers. Figure 7.18 illustrates two outliers.

7.8.2 Analysis of variance (ANOVA) table

The variance can be analyzed in detail by using a so-called *analysis of variance* (ANOVA) table. An ANOVA table can be used (i) to determine whether a model is significant and

Table 7.1. Analysis of variance (ANOVA) table

Source	SS	df	MS	F_{obs}
Regression	SS_{Reg}	v_{Reg}	$MS_{Reg} = \dfrac{SS_{Reg}}{v_{Reg}}$	$\dfrac{MS_{Reg}}{s^2}$
Residual	SSE	v	$s^2 = \dfrac{SSE}{v}$	
Lack of fit	SS_{LoF}	v_{LoF}	$MS_{LoF} = \dfrac{SS_{LoF}}{v_{LoF}}$	$\dfrac{MS_{LoF}}{MS_{PE}}$
Pure error	SS_{PE}	v_{PE}	$MS_{PE} = \dfrac{SS_{PE}}{v_{PE}}$	
Total, corrected for mean	$SS_{tot,\,corr}$	$v_{tot,corr}$		

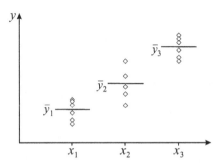

Figure 7.19. Data for three factors, showing differences within and between samples.

(ii) to check on its lack of fit. Repeated experiments are needed to examine this. Data for three different factors (x_1, x_2, and x_3) are shown in Figure 7.19. Each experiment is repeated multiple times in order to determine its variance. The figure shows that the variance is larger for x_2 than for the others because its points are more spread out. The mean value for y, $\bar{y}$, is shown for the three factors. With the use of the ANOVA table, we can investigate if there is a lack of fit to the model. We can also examine whether the differences between $\bar{y}_i$ are significant or are due to the variance of the points.

The ANOVA table is shown in Table 7.1; and in the following paragraph different terms in the table and its use are discussed.

The first two lines represent the regression model and the residual, where the residual can be divided into two parts: lack of fit and pure error. We start with the regression and residual used to test if the model is significant. The sum of squares due to regression, SS_{Reg}, can be calculated according to

$$SS_{Reg} = \sum_{i=1}^{n} (\hat{y}_i - \bar{y})^2. \tag{7.97}$$

This value is corrected to the mean, the background to subtracting the number of parameters by one (for the mean) to retrieve the degrees of freedom:

$$v_{Reg} = p - 1, \tag{7.98}$$

where p represents the number of parameters. The sum of square errors, SSE, can be calculated as described earlier as

$$SSE = \sum_{i=1}^{n} \varepsilon_i^2 = \sum_{i=1}^{n} (y_i - \hat{y}_i)^2, \tag{7.99}$$

with

$$v = n - p \tag{7.100}$$

degrees of freedom and

$$n = \sum_{j=1}^{m} n_j, \tag{7.101}$$

where n represents the total number of observations and m is the number of different x_i. For example, in Figure 7.19, $m = 3$ and $n = 6 + 5 + 6 = 17$. The sum of square errors contains both the errors associated with the pure error, SS_{PE}, and lack of fit SS_{LoF}, i.e.

$$SSE = SS_{PE} + SS_{LoF}. \tag{7.102}$$

The sum of squares of the pure error can be calculated from repeated identical experiments, according to

$$SS_{PE} = \sum_{j=1}^{m} \sum_{i=1}^{n_j} (y_{ji} - \bar{y}_j)^2, \tag{7.103}$$

with the number of degrees of freedom given by

$$v_{PE} = \sum_{j=1}^{m} (n_j - 1) = n - m. \tag{7.104}$$

The lack of fit of the model can be calculated via

$$SS_{LoF} = \sum_{j=1}^{m} n_j (\hat{y}_j - \bar{y}_j)^2, \tag{7.105}$$

with

$$v_{LoF} = m - p \tag{7.106}$$

degrees of freedom. However, an easier way to determine SS_{LoF} is to calculate it based on Equation (7.102), resulting in

$$SS_{LoF} = SSE - SS_{PE}. \tag{7.107}$$

Here, it becomes clear that the number of degrees of freedom for the lack of fit may be calculated from v and v_{PE}:

$$v_{LoF} = v - v_{PE} = n - p - (n - m) = m - p. \tag{7.108}$$

The total corrected sum of squares is given by

$$SS_{tot,corr} = \sum_{i=1}^{n} (y_i - \bar{y})^2. \tag{7.109}$$

It can also be calculated by adding the sum of squares for the regression and residual, which results in

$$SS_{tot,corr} = SS_{Reg} + SSE. \tag{7.110}$$

It is referred to as "corrected" because it is corrected by the total mean, and this correction is done in SS_{Reg}. Simultaneously, the degrees of freedom are corrected by -1 (due to the mean), so

$$v_{tot,corr} = n - 1. \tag{7.111}$$

Dividing the sum of squares by the number of degrees of freedom for the respective case yields the values for the mean squares, MS_{Reg}, s^2, MS_{LoF}, MS_{PE}, $MS_{tot,corr}$. The pure error mean square, MS_{PE}, is an estimate of σ^2 that is valid both when there is a lack of fit in the model and when there isn't.

The reason for setting up the ANOVA table is to determine if the model is significant and if there is any lack of fit to the model. First, we must determine if there is a lack of fit to the model. If there is, we cannot use s^2 as an estimation of σ^2 and thus cannot check the significance of the model. The lack of fit can be determined by examining the observed F-value for the lack of fit, that is

$$F_{obs,LoF} \simeq \frac{MS_{LoF}}{MS_{PE}}. \tag{7.112}$$

This value should be compared to the F-value from the F-distribution with v_{LoF} and v_{PE} degrees of freedom at α confidence level (for example, at the 95% or the 99% level), which is denoted $F(v_{LoF}, v_{PE}; \alpha)$. In the F-table (see Appendix C), data are shown for $F(v_1, v_2; \alpha)$; thus $v_1 = v_{LoF}$ and $v_2 = v_{PE}$. If $F_{obs,LoF} > F(v_{LoF}, v_{PE}; \alpha)$ is valid, the lack of fit is significant. For this case, it is not possible to determine confidence regions, significance of the model, etc. Instead, residual plots and further examination of the model are needed in order to uncover the reason for the lack of fit and thereafter to improve the model until there is no longer any lack of fit. For the case when $F_{obs,LoF} < F(v_{LoF}, v_{PE}; \alpha)$, there is no lack of fit observed and it becomes possible to determine the significance of the model. In this case, we can use s^2 as an estimation of σ^2. However, the fact that there is no lack of fit does not mean that the model is perfect, it merely means that we do not observe any lack of fit in these analyses.

In order to examine the significance of the model, the F-value observed for the regression, $F_{obs,Reg}$, should be calculated:

$$F_{obs,Reg} = \frac{MS_{Reg}}{s^2}. \tag{7.113}$$

This value should be compared to $F(v_{Reg}, v; \alpha)$ and if $F_{obs,Reg} > F(v_{Reg}, v; \alpha)$ is valid, the model is significant at the α level. However, if the opposite holds true, i.e. $F_{obs,Reg} < F(v_{Reg}, v; \alpha)$, the model is not significant, and it is recommended not to use this model.

7.8.3 R^2 statistic

An important tool for examining the model is to use the R^2 statistic, which is calculated by

$$R^2 = \frac{SS_{Reg}}{SS_{tot,corr}} = \frac{\sum_{i=1}^{n} (\hat{y}_i - \bar{y})^2}{\sum_{i=1}^{n} (y_i - \bar{y})^2}. \tag{7.114}$$

Thus, R^2 represents the fraction of the total variation around the mean, $\bar{y}$, explained by the model. Often, R^2 is multiplied by 100 and expressed in percentage terms. Generally, R^2 can assume values between 0 and 1. However, because the model can never explain the pure error, R^2 cannot be 1. The higher the value of R^2, the more capable the model is of describing experimental data. There are, however, also some issues with using R^2. If, for example, there are five different experimental points and you use five parameters in your model, you will receive a model that goes through all five points and thereby gives $R^2 = 1$. Of course, this does not mean that you have a good model, only that the model is over-parameterized, for the set of performed experiments. For this example, it is obvious that too many parameters were chosen. However, it is also important to examine this when repeating experiments, where it might not be obvious when the model is over-parameterized. Assume, for example, that you have 80 experimental points, and that there are 20 repeated experiments for each x point. If you choose a model with four parameters, you will get a perfect fit and a very high R^2 value, but this model is over-parameterized or is developed with too little variation in the experimental conditions.

7.9   Case study 7.1: Statistical analysis of a linear model

In an article about the growth rates of ice crystals (Ryan, B. F., Wishart, E. R., and Show, D. E. (*J. Atmos. Sci.* **33**, 842–850, 1976)), experiments were conducted in which ice crystals were placed into a compartment at a constant temperature (–5 °C). In order to analyze the growth of the ice crystals as a function of time, the saturation of the air by water was kept constant. The experimental data points were randomized over time. The experimental data are presented in Table 7.2, where y is the axial length of the crystals

Table 7.2. Data for Case study 7.1

x (s)	y (μm)	x (s)	y (μm)
50	19	125	28
60	20, 21	130	31, 32
70	17, 22	135	34, 25
80	25, 28	140	26, 33
90	21, 25, 31	145	31
95	25	150	36, 33
100	30, 29, 33	155	41, 33
105	35, 32	160	40, 30, 37
110	30, 28, 30	165	32
115	31, 36, 30	170	35
120	36, 25, 28	180	38

in microns and x is the time in seconds. Repeated measurements were also performed in order to examine the lack of fit.

Use a straight-line model, $y = \beta_0 + \beta_1 x + \varepsilon$, and fit it to the data. In addition, make a complete statistical analysis of the model. Use a 95% confidence level, i.e. $\alpha = 0.05$.

7.9.1   Solution

The experimental data are presented in vector and matrix notation:

$$
y = \begin{bmatrix} 19 \\ 20 \\ 21 \\ 17 \\ \cdot \\ \cdot \\ \cdot \\ 35 \\ 38 \end{bmatrix} \quad X = \begin{bmatrix} 1 & 50 \\ 1 & 60 \\ 1 & 60 \\ 1 & 70 \\ \cdot & \cdot \\ \cdot & \cdot \\ \cdot & \cdot \\ 1 & 170 \\ 1 & 180 \end{bmatrix}
$$

where y is an $n \times 1$ vector and X is an $n \times p$ matrix; $n = 43$ is the total number of experimental points, and p represents the number of parameters, in this case two (b_0 and b_1). The first column in X should only contain 1 at each position. The number of different x-values is 22, which is denoted m ($m = 22$). As there are many repeated experiments, m is significantly lower than the total number of experimental points (n). The number of observations for each x_i is named n_i, where $i = 1, \ldots, m$ and $n = \sum_{i=1}^{m} n_i$.

The model parameters can be calculated as follows:

$$
b = \begin{bmatrix} b_0 \\ b_1 \end{bmatrix} = (X'X)^{-1}X'y = \begin{bmatrix} 0.33355 & -0.0026342 \\ -0.0026342 & 2.23641 \times 10^{-5} \end{bmatrix} \begin{bmatrix} 1292 \\ 158205 \end{bmatrix} = \begin{bmatrix} 14.19 \\ 0.1346 \end{bmatrix}.
$$

Thus, the following model is obtained:

$$
\hat{y} = 14.19 + 0.1346x.
$$

7.9.1.1   Evaluation of the model

When starting to evaluate the developed model, it is advisable to perform both regression plots and residual plots. Figure 7.20 depicts a regression plot. Although the variance is high, the data follow the linear model closely. There are no clear trends or evidence of outliers.

Figures 7.21 and 7.22 show the residual and Studentized residual versus the response predicted, respectively. These plots show that the residuals are evenly distributed and that there are no clear trends. In the Studentized residual plot (Figure 7.22), all points fall between -3 and 3, with no evidence of outliers (see Section 7.8.1).

As the residual plots evidence good behavior, we will proceed to assemble an ANOVA table (Table 7.3) in order to study the lack of fit and the significance of the model (see Section 7.8.2).

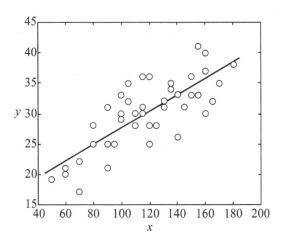

Figure 7.20. Regression plot.

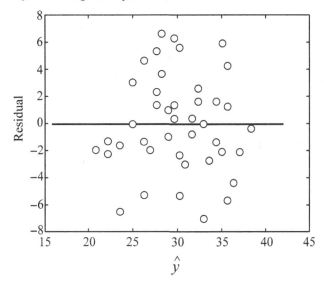

Figure 7.21. Residual plot.

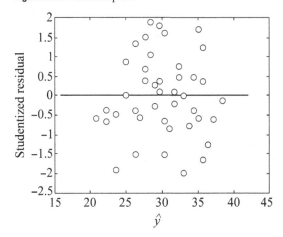

Figure 7.22. Studentized residual plot.

Table 7.3. ANOVA table for Case study 7.1

Source	SS	df	MS	F_{obs}
Regression	810.3	$p - 1 = 2 - 1 = 1$	810.3	63.45
Residual	523.6	$n - p = 43 - 2 = 41$	12.77	
Lack of fit	239.6	$m - p = 22 - 2 = 20$	11.98	0.886
Pure error	284	$n - m = 43 - 22 = 21$	13.52	
Total, corrected for the mean	1334	$n - 1 = 43 - 1 = 42$		

7.9.1.2   Lack of fit

The lack of fit is examined by comparing F_{obs} to the F-value from the F-table (see Appendix C), i.e.

$$F(v_{LoF}, v_{PE}; \alpha) = F(20, 21; 0.05) = 2.1,$$

where the calculation for the degrees of freedom (df), for lack of fit (v_{LoF}), and pure error (v_{PE}) are found in Table 7.3.

According to the calculation in the ANOVA table, $F_{obs,LoF} = 0.886$, which is smaller than $F(v_{LoF}, v_{PE}; \alpha) = 2.1$. Thus, the lack of fit is not significant and there is no reason to doubt this model. We will therefore examine whether or not the regression model is significant.

According to Table 7.3, the significance of the model is investigated by calculating $F_{obs,Reg} = MS_{Reg}/s^2 = 63.45$. This value is compared with $F(v_{Reg}, v; \alpha)$, which becomes $F(1; 41; 0.05) = 4.08$ (see Table 7.3 for the calculation of the degrees of freedom). Thus, $F_{obs,Reg} > F(v_{Reg}, v; \alpha)$, which means that the model is significant at the α level. In addition, the R^2 statistic is determined according to Equation (7.114), i.e.

$$R^2 = \frac{SS_{Reg}}{SS_{tot,corr}} = 0.608.$$

Thus, the regression model describes a great deal of the variation in the data.

7.9.1.3   Confidence intervals

The calculations of the confidence intervals are described in Section 7.5. Briefly, the $1 - \alpha$ confidence interval for the parameter b_p in the model is given by

$$b_p \pm se(b_p)t(n - p; \alpha/2),$$

where

$$se(b_p) = s\sqrt{\{(X'X)^{-1}\}_{pp}};$$

s^2 is already calculated and presented in the ANOVA table (Table 7.3). The confidence intervals are presented in Table 7.4.

The confidence interval can also be written using the following notation:

$$10.02 \le b_0 \le 18.36;$$

$$0.1005 \le b_1 \le 0.1687.$$

Table 7.4. Confidence intervals for Case study 7.1

Parameter	b_p	$se(b_p)$	t_{obs}	Confidence interval
b_0	14.19	2.064	6.875	14.19 ± 4.17
b_1	0.1346	0.0169	7.966	0.1346 ± 0.0341

Student's t-tests can be used to determine the significance of the parameters. In order to do so, the t-value from the t-distribution is needed (see Appendix C):

$$t(n - p; \alpha/2) = t(43 - 2; 0; 0.05/2) = 2.02.$$

This value should be compared with the calculated $t_{obs} = b_p/se(b_p)$, which are shown in Table 7.4. For both parameters $t_{obs} > 2.02$, thus both parameters in the model are significant.

The confidence interval for the mean response, $\mu(x_0)$, at $x_0 = 100$, can be calculated according to

$$\hat{y}(x_0) \pm s \sqrt{x_0'(X'X)^{-1}x_0} \, t(n - p; \alpha/2),$$

resulting in

$$x_0 = 27.65 \pm 1.26,$$
$$26.4 \le \mu(x_0) \le 28.9.$$

7.9.1.4   Correlation

The correlation coefficient can be determined by

$$C_{01} = \frac{\{(X'X)^{-1}\}_{01}}{\sqrt{\{(X'X)^{-1}\}_{00}\{(X'X)^{-1}\}_{11}}} = -0.9645.$$

Thus the correlation is quite high.

7.10   Case study 7.2: Multiple regression

A die-casting process was examined and experiments conducted in order to study the effects of the furnace temperature (x_1) and the die-closing time (x_2) on the temperature difference of the die surface (y). The data are shown in Table 7.5.

Fit the model $y = \beta_0 + \beta_1 x_1 + \beta_2 x_2 + \varepsilon$ to the data. In addition, make a complete statistical analysis of the model. Use a 95% confidence level, i.e. $\alpha = 0.05$.

Table 7.5. Data for Case study 7.2

x_1	x_2	y
1250	6	80
1300	7	95
1350	6	101
1250	7	85
1300	6	92
1250	8	87
1300	8	96
1350	7	106
1350	8	108

See Chang, S. I. and Shivpuri, R., *Qual. Engng.* **7**, 371–383, 1994.

7.10.1   Solution

The experimental data are presented below in vector and matrix notation:

$$
y = \begin{bmatrix} 80 \\ 95 \\ 101 \\ 85 \\ 92 \\ 87 \\ 96 \\ 106 \\ 108 \end{bmatrix} ; \quad
X = \begin{bmatrix} 1 & 1250 & 6 \\ 1 & 1300 & 7 \\ 1 & 1350 & 6 \\ 1 & 1250 & 7 \\ 1 & 1300 & 6 \\ 1 & 1250 & 8 \\ 1 & 1300 & 8 \\ 1 & 1350 & 7 \\ 1 & 1350 & 8 \end{bmatrix} ,
$$

where y is an $n \times 1$ vector and X is an $n \times p$ matrix; $n = 9$ represents the total number of experimental points, and p represents the number of parameters, in this case three (b_0, b_1, and b_2). The first column of X should only contain 1 at each position. In these experiments, there are no repeated experimental points, and therefore the lack of fit cannot be determined.

The model parameters are given by:

$$
b = \begin{bmatrix} b_0 \\ b_1 \\ b_2 \end{bmatrix} = (X'X)^{-1} X' y
$$

$$
= \begin{bmatrix} 120.94 & -0.086667 & -1.1667 \\ -0.086667 & 6.6667 \times 10^{-5} & 5.1283 \times 10^{-17} \\ -1.1667 & 5.1283 \times 10^{-17} & 0.16667 \end{bmatrix} \begin{bmatrix} 850 \\ 1108150 \\ 5968 \end{bmatrix} = \begin{bmatrix} -199.56 \\ 0.21 \\ 3.00 \end{bmatrix} .
$$

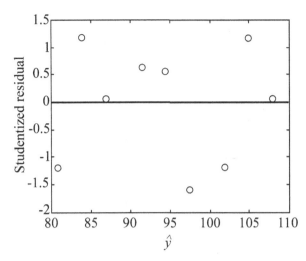

Figure 7.23. Studentized residual plot versus $\hat{y}$.

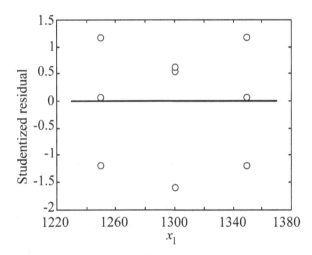

Figure 7.24. Studentized residual plot versus x_1.

The following model is obtained:

$$\hat{y} = -199.56 + 0.21x_1 + 3.00x_2.$$

7.10.1.1 Evaluation of the model

First, we start by examining the residual plots in order to ensure that there are no clear trends for the residuals. Such trends would mean that the model would not be correctly formulated. Figures 7.23, 7.24, and 7.25 show the Studentized residuals versus the predicted response ($\hat{y}$), x_1 and x_2, respectively. These plots show that the residuals

Table 7.6. ANOVA table for Case study 7.2

Source	SS	df	MS	F_{obs}
Regression	715.5	$p - 1 = 3 - 1 = 2$	357.8	319.3
Residual	6.72	$n - p = 9 - 3 = 6$	1.12	
Total, corrected for the mean	722.2	$n - 1 = 9 - 1 = 8$		

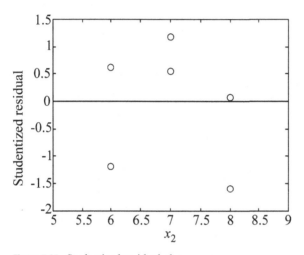

Figure 7.25. Studentized residual plot versus x_2.

are evenly distributed and that there are no clear trends. In addition, all points fall between -3 and 3, with no evidence of outliers (see Section 7.8.1).

Since the residual plots evidence good results, it is now possible to determine the significance of the model. See the ANOVA table (Table 7.6).

The significance of the model is examined by calculating $F_{obs,Reg} = MS_{Reg}/s^2 = 319.3$ (see Table 7.6). The F-value can be retrieved from the F-table (see Appendix C) and $F(v_{Reg}, v; \alpha)$, which for this case study is $F(2, 6; 0.05) = 5.14$. This results in $F_{obs,Reg} > F(v_{Reg}, v; \alpha)$, i.e. the model is significant at the α level. Further, the R^2 statistic is determined using Equation (7.114) to be

$$R^2 = \frac{SS_{Reg}}{SS_{tot,corr}} = 0.991.$$

Thus, we observe that the regression model can describe a large fraction of the variation in the data.

7.10.1.2   Confidence intervals

The calculations of the confidence intervals are discussed in Section 7.5. The $1 - \alpha$ confidence interval for the parameter b_p in the model can be determined from

$$b_p \pm se(b_p)t(n - p; \alpha/2),$$

Table 7.7. Confidence intervals for Case study 7.2

Parameter	b_p	$se(b_p)$	t_{obs}	Confidence interval
b_0	-199.6	11.64	-17.14	-199.6 ± 28.5
b_1	0.21	0.00864	24.30	0.21 ± 0.021
b_2	3.00	0.432	6.94	3.00 ± 1.06

where

$$se(b_p) = s\sqrt{\{(X'X)^{-1}\}_{pp}};$$

s^2 is given in Table 7.6.

The confidence intervals for the parameters are shown in Table 7.7. The confidence intervals can also be expressed as follows:

$$-228.04 \leq b_0 \leq -171.07,$$

$$0.1889 \leq b_1 \leq 0.2311,$$

$$1.94 \leq b_2 \leq 4.06.$$

Student's t-tests can be used for determining the significance of the parameters. A t-value is taken from the t-distribution (see Appendix C):

$$t(n - p; \alpha/2) = 2.45.$$

The term t_{obs} shown in Table 7.7 is calculated from $b_p/se(b_p)$. These values are then compared with $t(n - p; \alpha/2)$. For all parameters, $t_{obs} > 2.45$, thus b_0, b_1, and b_2 are significant. Note that t_{obs} for b_0 is negative. However, the absolute value for t_{obs} should be used in these tests, yielding 17.14 for b_0.

The confidence interval for the mean response, $\mu(x_0)$, at $x_0' = [1 \ 1250 \ 8]$, can be calculated according to

$$\hat{y}(x_0) \pm s\sqrt{x_0'(X'X)^{-1}x_0} \, t(n - p; \alpha/2)$$

which results in

$$\mu(x_0) = 86.94 \pm 1.73,$$

$$85.2 \leq \mu(x_0) \leq 88.7.$$

The confidence regions are given by

$$(\beta - b)'X'X(\beta - b) \leq ps^2 F(p, n - p; \alpha).$$

The results for the confidence region and interval are shown in Figure 7.26 for b_1 and b_2, while keeping b_0 constant. Note that the scales are different on the x- and y-axes. Thus the confidence interval and band are much larger for b_2.

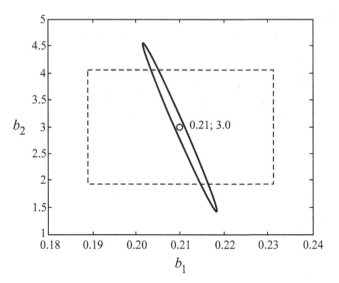

Figure 7.26. Confidence region and confidence intervals. Note that the scales are different on the axes.

From Figure 7.26, it is clear that there is a small correlation between the parameters b_1 and b_2. In addition, it becomes evident that the confidence region is very different from the confidence intervals.

7.10.1.3 Correlation

The correlation matrix can be determined by

$$C_{ij} = \frac{\{(X'X)^{-1}\}_{ij}}{\sqrt{\{(X'X)^{-1}\}_{ii}\{(X'X)^{-1}\}_{jj}}},$$

which produces

$$C = \begin{bmatrix} 1 & -0.9652 & -0.2599 \\ -0.9652 & 1 & 1.5 \times 10^{-14} \\ -0.2599 & 1.5 \times 10^{-14} & 1 \end{bmatrix}.$$

Note the symmetry of the matrix due to the fact that the correlation between b_1 and b_2 is the same as that between b_2 and b_1. The results in the matrix show that the correlation is high between b_0 and b_1. However, the correlation between b_0 and b_2 is low, and that between b_1 and b_2 is extremely low.

7.11 Case study 7.3: Non-linear model with one predictor

Hydrogen peroxide has been used to oxidize a secondary alcohol in a stirred batch reactor. In order to increase the reaction rate, a solid catalyst was added. The conversion (y) of hydrogen peroxide was detected at different times (t); see Table 7.8.

Table 7.8. Data for Case study 7.3

t (min)	y
20	0.0476
50	0.0700
80	0.129
110	0.152
140	0.126
170	0.157
200	0.173
230	0.140
260	0.210
290	0.155
320	0.197

Table 7.9. ANOVA table for Case study 7.3

Source	SS	df	MS	F_{obs}
Regression	0.240	$p = 2$	0.12	244
Residual	0.0044	$n - p = 11 - 2 = 9$	0.00049	
Total, corrected for the mean	0.244	$n = 11$		

Fit the non-linear model $y = \theta_1(1 - e^{-\theta_2 t})$ to the data. Thereafter examine the significance of the model and the parameter significance. In addition, construct confidence bands and joint confidence regions. Use a 95% confidence level, i.e. $\alpha = 0.05$.

7.11.1   Solution

The non-linear model parameters are determined numerically using the least square method in MATLAB. In order achieve this, the sum of square errors is used according to

$$SSE(\theta) = \sum_{i=1}^{n} \varepsilon_i^2 = \sum_{i=1}^{n} (y_i - f(x_i, \hat{\theta}))^2.$$

This results in the following parameters and model:

$$\theta_1 = 0.183, \qquad \theta_2 = 0.0122, \qquad \text{and} \qquad y = 0.183(1 - e^{-0.0122 t}).$$

7.11.1.1   ANOVA table

The ANOVA table is presented in Table 7.9. The reason that the degrees of freedom equal p not $p - 1$ is that this model does not contain a constant due to the mean (θ_0).

The significance of the model is examined by comparing $F_{obs,Reg}$ with $F(v_{Reg}, v; \alpha)$, which can be retrieved from the F-table. For this example, $F(2, 9; 0.05) = 4.26$ at the 95% level. This results in $F_{obs,Reg} > F(v_{Reg}, v; \alpha)$, thus the model is significant at the α level.

Table 7.10. Confidence intervals for Case study 7.3

Parameter	$\hat{\theta}_p$	$se(\hat{\theta}_p)$	t_{obs}	Confidence interval
$\hat{\theta}_1$	0.183	0.0147	12.43	0.183 ± 0.0333
$\hat{\theta}_2$	0.0122	0.00319	3.821	0.0122 ± 0.00723

7.11.1.2 Confidence intervals and bands

The $1 - \alpha$ confidence interval for the parameter $\hat{\theta}_p$ in the model can be determined according to

$$\hat{\theta}_p \pm se(\hat{\theta}_p)t(n - p; \alpha/2),$$

where

$$se(\hat{\theta}_p) = s\sqrt{\{(\boldsymbol{J'J})^{-1}\}_{pp}},$$

where $\boldsymbol{J}$ is the Jacobian and can be calculated according to

$$J_{n,p} = \left[\frac{\partial f(\boldsymbol{x_n}, \hat{\boldsymbol{\theta}})}{\partial \theta_p}\right]_{\theta = \hat{\theta}} \approx \frac{f(\boldsymbol{x_n}, \hat{\boldsymbol{\theta}}) - (\boldsymbol{x_n}, \theta_{p,1\%})}{\Delta \theta_p}$$

or directly through the software. In this case, MATLAB's built-in functions have been used to generate the Jacobian, resulting in

$$\boldsymbol{J}(\hat{\theta}_1, \hat{\theta}_2) = \begin{bmatrix} 0.2166 & 2.8649 \\ 0.4568 & 4.9662 \\ 0.6234 & 5.5096 \\ 0.7388 & 5.2529 \\ 0.8189 & 4.6356 \\ 0.8744 & 3.9030 \\ 0.9129 & 3.1839 \\ 0.9396 & 2.5388 \\ 0.9581 & 1.9900 \\ 0.9710 & 1.5390 \\ 0.9799 & 1.1775 \end{bmatrix},$$

which yields

$$\boldsymbol{J'J} = \begin{bmatrix} 7.1627 & 27.2607 \\ 27.2607 & 151.8384 \end{bmatrix}$$

and

$$(\boldsymbol{J'J})^{-1} = \begin{bmatrix} 0.4408 & -0.0791 \\ -0.0791 & 0.0208 \end{bmatrix};$$

$s^2 (= 0.00049)$ is given in Table 7.9. The calculations for the confidence intervals are shown in Table 7.10.

The significance of the parameters is evaluated using the t-test. The t-value $t(n - p; \alpha/2)$ is 2.26 (for the t-distribution, see Appendix C). Since the values of t_{obs} for

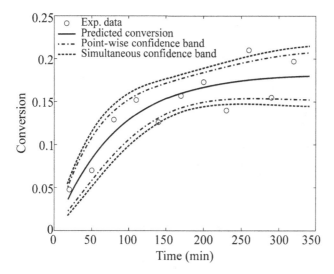

Figure 7.27. Experimental data together with predicted conversion. In addition, the point-wise and simultneous confidence bands are presented.

both parameters are larger than $t(n - p; \alpha/2)$, $\hat{\theta}_1$ and $\hat{\theta}_2$ are significant; t_{obs} is calculated using $t_{obs} = \hat{\theta}_p/se(\hat{\theta}_p)$.

The $1 - \alpha$ confidence interval for the dependent variable $f(x_0, \boldsymbol{\theta})$, which can also be called the "mean response," is retrieved by

$$f(x_0, \boldsymbol{\theta}) \pm s \sqrt{\boldsymbol{j}_0'(\boldsymbol{J}'\boldsymbol{J})^{-1}\boldsymbol{j}_0} \; t(n - p; \alpha/2),$$

where

$$\boldsymbol{j}_0 = \left[\frac{\partial f(x_0, \boldsymbol{\theta})}{\partial \boldsymbol{\theta}} \right]_{\boldsymbol{\theta} = \hat{\boldsymbol{\theta}}}.$$

The confidence intervals for several points, calculated using the preceding equation, together form a so-called "$1 - \alpha$ point-wise confidence band." The "simultaneous confidence band" at the $1 - \alpha$ level is determined by

$$f(x, \boldsymbol{\theta}) \pm s \sqrt{\boldsymbol{j}'(\boldsymbol{J}'\boldsymbol{J})^{-1}\boldsymbol{j}} \sqrt{pF(p; n - p; \alpha)}.$$

The experimental data are shown in Figure 7.27 together with the predicted conversion using the developed model. In addition, the resulting point-wise and simultaneous confidence bands are shown. The results show that the point-wise and simultaneous confidence bands are similar for this example.

7.11.1.3 Joint confidence regions

The approximate elliptical contour, with exact probability level $1 - \alpha$, is calculated using

$$(\boldsymbol{\theta} - \hat{\boldsymbol{\theta}})' \times \boldsymbol{J}'\boldsymbol{J} \times (\boldsymbol{\theta} - \hat{\boldsymbol{\theta}}) \leq ps^2 F(p, n - p, 1 - \alpha),$$

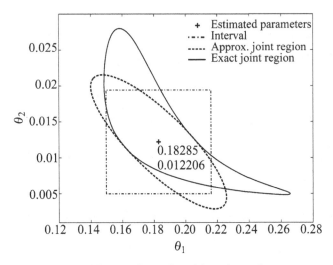

Figure 7.28. Confidence region and confidence intervals.

which yields

$$[\theta_1 - 0.1829 \;\; \theta_2 - 0.0122] \times \boldsymbol{J'J} \times \begin{bmatrix} \theta_1 - 0.1829 \\ \theta_2 - 0.0122 \end{bmatrix} \leq 4.177 \cdot 10^{-3}.$$

The exact contour with approximate probability level $1 - \alpha$ is determined using

$$S(\boldsymbol{\theta}) - S(\hat{\boldsymbol{\theta}}) \leq ps^2 F(p, n - p, 1 - \alpha),$$

where

$$S(\hat{\boldsymbol{\theta}}) = SSE.$$

The results for the confidence regions are shown in Figure 7.28, together with the individual confidence intervals for $\hat{\theta}_1$ and $\hat{\theta}_2$. The results clearly show that there is a large difference between the individual confidence intervals and the joint confidence regions, but also that there is a difference between the elliptical contour and the exact region. Further, the results of the joint confidence region also show that there is a correlation between the parameters θ_1 and θ_2.

7.11.1.4 Correlation
The correlation is determined using the correlation matrix according to

$$C_{ij} = \frac{\{(\boldsymbol{J'J})^{-1}\}}{\sqrt{\{(\boldsymbol{J'J})^{-1}\}_{ii}\{(\boldsymbol{J'J})^{-1}\}_{jj}}},$$

which gives

$$C = \begin{bmatrix} 1 & -0.8266 \\ -0.8266 & 1 \end{bmatrix}.$$

Thus, there is a quite high correlation between the parameters.

7.12 Questions

(1) Set up the normal equations and derive expressions for b_0 and b_1 in a linear model.

(2) How can the parameters in vector b be determined using the matrix form?

(3) Why are weighted residuals used?

(4) What is an outlier?

(5) How do you determine confidence intervals and regions, respectively?

(6) What are the differences between confidence intervals and confidence bands?

(7) Describe the ANOVA table and how to calculate the different values in the table.

(8) What is the correlation between parameters and how do you determine it?

(9) Why is the correlation matrix symmetrical?

(10) What makes a model non-linear?

(11) Give an example of an intrinsically linear model.

(12) How do you determine confidence intervals for a non-linear model?

(13) Give examples of residual plots when the residuals are evenly distributed and when there are clear trends. Also describe how the models should be changed when clear trends are discernible.

(14) How is R^2 determined in order to get information about how much of a variation can be described by a model?

7.13 Practice problems

7.1 The dependence of the reaction rate for the pressure of a hydrogenation reaction in a tubular reactor has empirically been shown to follow the model

$$r(P) = \frac{\theta_1 \cdot P}{1 + \theta_2 \cdot P}$$

where r is the reaction rate (mol s^{-1}, per kilogram of catalyst) and P is the total pressure (bar). At 170 °C, the reaction rates given in Table 7.11 were determined.

(a) Determine the parameters $\hat{\theta}_1$ and $\hat{\theta}_2$ by using one of MATLAB's (or that of another software) least squares functions to search for the parameter values that minimize the model error. In order to obtain a rapid convergence against the global minima, good start guesses of $\hat{\theta}_1$ and $\hat{\theta}_2$ are important.

(b) Determine the individual confidence intervals for the parameters at a 95% level.

Table 7.11. Reaction rates as a function of pressure

r (mol s^{-1} per kg catalyst)	P (bar)
0.0192	1.382
0.0234	2.075
0.0258	2.431
0.0281	2.762
0.0318	3.500
0.0319	3.833
0.0334	4.129

Table 7.12. Reaction rates for the two cases treated and untreated with pyromycin

	Reaction rate (counts min^{-2})	
Substrate concentration (ppm)	Treated with pyromycin	Untreated
0.025	79	69
0.025	45	55
0.07	94	87
0.07	111	81
0.105	127	101
0.105	142	111
0.23	163	129
0.23	156	127
0.54	194	142
0.54	199	159
1.18	209	162
1.18	205	161

(c) Plot the joined, approximate confidence region of the parameters at the 95% level. Hint: If you use MATLAB, the *contour* function can be a great help.

(d) Determine the prediction interval for a new observation at 2.8 bar pressure.

7.2 The possible effect the substance pyromycin might have on an enzymatic reaction is examined. Two experimental series are conducted, one applying the pyromycin treatment, the other without. In the tests, the dependence of the reaction rate on the substrate concentration is measured. The kinetics can be assumed to follow the Michaelis–Menten model:

$$f(\boldsymbol{\theta}, \boldsymbol{x}) = \frac{\theta_1 \cdot \boldsymbol{x}}{\theta_2 + \boldsymbol{x}}.$$

An hypothesis is that pyromycin influences parameter θ_1, not θ_2. Do a statistical analysis and, based on that result, determine whether the hypothesis is likely to be true or not. The experimental data are given in Table 7.12.

Table 7.13. Data for pressure drop, pressure, heat loading, and mass flow

Measured pressure drop (kPa m^{-1})	Calculated pressure drop (kPa m^{-1})	Absolute pressure (kPa)	Heat loading (kW m^{-2})	Mass flow (kg m^{-2} s^{-1})
1.440	1.188	296.9	3.64	316
2.174	1.731	296.0	19.35	201
1.193	0.758	292.9	14.65	125
1.602	1.585	594.7	3.24	316
2.545	2.849	595.8	18.96	201
1.204	1.251	594.0	6.42	125
0.989	1.093	792.6	3.35	316
2.765	3.072	795.6	7.56	201
0.791	0.984	784.4	11.55	125

(a) Determine the parameters $\hat{\theta}_1$ and $\hat{\theta}_2$ by non-linear regression.
(b) Determine the individual 95% confidence intervals of the parameters. Are the parameters correlated; if your answer is yes, how?
(c) Conduct a residual analysis using Studentized residuals.
(d) Determine if there are any significant model errors (lack of fit).
(e) Perform the corresponding analysis for the untreated series.
(f) Based on these results, what do you think of the stated hypothesis?

7.3 A PhD student at Chalmers University measured the pressure drop during evaporation of 1,1,1,2-tetrafluoroethane in tubes. The student also calculated what a certain model, in the form of pressure drop $= f$ (pressure, heat loading, mass flow), was capable of predicting for the same conditions. The conditions and results are shown in Table 7.13.

Examine, with the help of residual analysis, if there is reason to suspect that the model needs further development, and, if so, in what regard. Normalization or Studentization is not needed.

7.4 When conducting a lack of fit analysis, the following ratio:

$$\frac{SS_{LoF}/v_{LoF}}{SS_{PE}/v_{PE}},$$

is compared with the F-distribution value for $F(v_1, v_{PE}, \alpha)$. In Figure P7.1, the reaction rate is given as a function of reactant concentration. The points represent experimentally measured values, and the solid line represents a model with two parameters. There are 29 points in total. For the five highest concentrations, repeated experiments are taking place.

(a) Which contribution to SS_{Res} (the residual sum of squares), SS_{LoF}, respectively SS_{PE}, originates from the two points between the concentrations 0.4 and 0.6? Illustrate graphically the differences that form the basis for SS_{Res} and SS_{PE}.
(b) What is the F-distribution value at the 95% level for the example in Figure P7.1?
(c) What does it mean if the ratio becomes larger than the F-distribution value?

Table 7.14. Input data for the model

N	T	C	Y_{obs}	$Y_{calculated}$	Res
1	285	7	2.3	2.94	−0.64
2	254	7	3.2	3.50	−0.3
3	302	9	3.0	2.57	0.43
4	323	6	1.6	2.23	−0.63
5	314	10	2.4	2.41	−0.01
6	282	8.5	3.7	2.95	0.75
7	321	5.3	1.4	2.26	−0.86
8	302	10	3.5	2.61	0.89
9	324	11.5	2.7	2.26	0.44

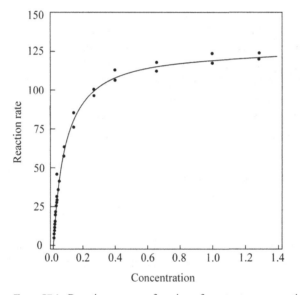

Figure P7.1. Reaction rate as a function of reactant concentration.

7.5 In order to reduce the shrinkage of an artificial fiber, it is pretreated by dipping into a solution with concentration C and then heat-treated at temperature T. The goal is to make a model describing how the shrinkage Y depends on the pretreatment.

Nine experiments were conducted; the data are shown in Table 7.14. The measured shrinkage is denoted Y_{obs}. The results of a first version of the model are also given in the column $Y_{calculated}$. Note that N is the notation for the experiment number, and the residual is denoted **Res**.

With the help of residual analysis, examine the suggested model and propose, with motivation, if and how the model might be improved.

7.6 As part of an assignment to find the optimal process conditions in a steam stripper, a simple model is needed. The model should describe how the water content in the

Table 7.15. Water content in the distillate

x (kg s^{-1})	y
0.263	0.34
0.294	0.32
0.321	0.24
0.367	0.20
0.429	0.16
0.608	0.11
1.33	0.06
13.6	0.04

distillate (y) depends on the steam consumption (x). As a basis for the model, the data in Table 7.15 were collected during the experiments.

The following model is suggested:

$$y = a \left(\ln \left(\frac{x}{b} \right) \right)^{-2/3},$$

where a and b are the parameters in the model.

By minimizing the residual sum of squares, SSE, the following data were obtained:

- a (parameter 1) = 0.1321
- b (parameter 2) = 0.211 kg s^{-1}
- $SSE = 0.003726$
- $J^T J = \begin{bmatrix} 20.3453 & 25.9254 \\ 25.9254 & 38.9143 \end{bmatrix}$
- $(J^T J)^{-1} = \begin{bmatrix} 0.3254 & -0.2168 \\ -0.2168 & 0.1701 \end{bmatrix}.$

(a) Optimal steam flow is expected to be about 0.58 kg s^{-1}. Calculate a 95% confidence interval, which will describe the uncertainty of the model for this flow.

(b) Others that have conducted similar experiments have found $a = 0.107$ and $b = 0.211$. Are these parameters within the confidence region on the 95% significance level?

Appendix A
Microscopic transport equations

A.1 Rectangular coordinates

Mass balance for component A (ρ and D_{AB} constant)

$$\frac{\partial C_A}{\partial t} + v_x \frac{\partial C_A}{\partial x} + v_y \frac{\partial C_A}{\partial y} + v_z \frac{\partial C_A}{\partial z} = D_{AB} \left(\frac{\partial^2 C_A}{\partial x^2} + \frac{\partial^2 C_A}{\partial y^2} + \frac{\partial^2 C_A}{\partial z^2} \right) + R_A.$$

Total mass balance (ρ constant)

$$\frac{\partial v_x}{\partial x} + \frac{\partial v_y}{\partial y} + \frac{\partial v_z}{\partial z} = 0.$$

Energy balance (ρ and k constant)

$$\rho c_p \left(\frac{\partial T}{\partial t} + v_x \frac{\partial T}{\partial x} + v_y \frac{\partial T}{\partial y} + v_z \frac{\partial T}{\partial z} \right) = k \left(\frac{\partial^2 T}{\partial x^2} + \frac{\partial^2 T}{\partial y^2} + \frac{\partial^2 T}{\partial z^2} \right) + S.$$

Momentum balance (ρ and μ constant)

$$x: \quad \rho \left(\frac{\partial v_x}{\partial t} + v_x \frac{\partial v_x}{\partial x} + v_y \frac{\partial v_x}{\partial y} + v_z \frac{\partial v_x}{\partial z} \right) = -\frac{\partial p}{\partial x} + \mu \left(\frac{\partial^2 v_x}{\partial x^2} + \frac{\partial^2 v_x}{\partial y^2} + \frac{\partial^2 v_x}{\partial z^2} \right)$$
$$+ \rho g_x,$$

$$y: \quad \rho \left(\frac{\partial v_y}{\partial t} + v_x \frac{\partial v_y}{\partial x} + v_y \frac{\partial v_y}{\partial y} + v_z \frac{\partial v_y}{\partial z} \right) = -\frac{\partial p}{\partial y} + \mu \left(\frac{\partial^2 v_y}{\partial x^2} + \frac{\partial^2 v_y}{\partial y^2} + \frac{\partial^2 v_y}{\partial z^2} \right)$$
$$+ \rho g_y,$$

$$z: \quad \rho \left(\frac{\partial v_z}{\partial t} + v_x \frac{\partial v_z}{\partial x} + v_y \frac{\partial v_z}{\partial y} + v_z \frac{\partial v_z}{\partial z} \right) = -\frac{\partial p}{\partial z} + \mu \left(\frac{\partial^2 v_z}{\partial x^2} + \frac{\partial^2 v_z}{\partial y^2} + \frac{\partial^2 v_z}{\partial z^2} \right)$$
$$+ \rho g_z.$$

A.2 Cylindrical coordinates (mass and energy)

Mass balance for component A (ρ and D_{AB} constant)

$$\frac{\partial C_A}{\partial t} + v_r \frac{\partial C_A}{\partial r} + v_\theta \frac{1}{r} \frac{\partial C_A}{\partial \theta} + v_z \frac{\partial C_A}{\partial z} = D_{AB} \left(\frac{1}{r} \frac{\partial}{\partial r} r \frac{\partial C_A}{\partial r} + \frac{1}{r^2} \frac{\partial^2 C_A}{\partial \theta^2} + \frac{\partial^2 C_A}{\partial z^2} \right)$$
$$+ R_A.$$

Energy balance (ρ and k constant)

$$\rho c_p \left(\frac{\partial T}{\partial t} + v_r \frac{\partial T}{\partial r} + v_\theta \frac{1}{r} \frac{\partial T}{\partial \theta} + v_z \frac{\partial T}{\partial z} \right) = k \left(\frac{1}{r} \frac{\partial}{\partial r} r \frac{\partial T}{\partial r} + \frac{1}{r^2} \frac{\partial^2 T}{\partial \theta^2} + \frac{\partial^2 T}{\partial z^2} \right) + S.$$

A.3   Spherical coordinates (mass and energy)

Mass balance for component A (ρ and D_{AB} constant)

$$\frac{\partial C_A}{\partial t} + v_r \frac{\partial C_A}{\partial r} + v_\theta \frac{1}{r} \frac{\partial C_A}{\partial \theta} + v_\phi \frac{1}{r \sin \theta} \frac{\partial C_A}{\partial \phi}$$

$$= D_{AB} \left(\frac{1}{r^2} \frac{\partial}{\partial r} r^2 \frac{\partial C_A}{\partial r} + \frac{1}{r^2 \sin \theta} \frac{\partial}{\partial \theta} \sin \theta \frac{\partial C_A}{\partial \theta} + \frac{1}{r^2 \sin^2 \theta} \frac{\partial^2 C_A}{\partial \phi^2} \right) + R_A.$$

Energy balance (ρ and k constant)

$$\rho c_p \left(\frac{\partial T}{\partial t} + v_r \frac{\partial T}{\partial r} + v_\theta \frac{1}{r} \frac{\partial T}{\partial \theta} + v_\phi \frac{1}{r \sin \theta} \frac{\partial T}{\partial \phi} \right)$$

$$= k \left(\frac{1}{r^2} \frac{\partial}{\partial r} r^2 \frac{\partial T}{\partial r} + \frac{1}{r^2 \sin \theta} \frac{\partial}{\partial \theta} \sin \theta \frac{\partial T}{\partial \theta} + \frac{1}{r^2 \sin^2 \theta} \frac{\partial^2 T}{\partial \phi^2} \right) + S.$$

Appendix B
Dimensionless variables

Useful dimensionless numbers can be created by taking the ratios of different forces, mass fluxes, or heat fluxes. The dimensionless number will indicate which of the forces or fluxes is the most important. Various dimensionless quantities are given in Tables B.1–B.4.

Table B.1. Dimensionless numbers based on force

	Inertial	Viscous	Buoyancy	Surface tension	Pressure
Force (kg m s^{-2})	$\rho U^2 L^2$	$\mu U L$	$gL^3 \Delta \rho$	γL	$L^2 \Delta p$
Inertial	–	$Re = \dfrac{\rho U L}{\mu}$	$Fr = \dfrac{gL\Delta\rho}{\rho U^2}$	$We = \dfrac{\rho U^2 L}{\gamma}$	$Eu = \dfrac{\Delta p}{\rho U^2}$
Viscous		–	$\dfrac{gL^2\Delta\rho}{\mu U}$	$Ca = \dfrac{\mu U}{\gamma}$	$\dfrac{L\Delta p}{\mu U}$
Buoyancy			–	$Eo = \dfrac{gL^2\Delta\rho}{\gamma}$	$\dfrac{\Delta p}{gL\Delta\rho}$
Surface tension				–	$\dfrac{L\Delta p}{\gamma}$
Pressure					–

Table B.2. Dimensionless numbers based on mass flux

	Convective	Diffusion	Interface	Reaction
Mass flux (mol s^{-1})	$U\Delta C L^2$	$D\Delta C L$	$k_w \Delta C L^2$	$r_A L^3$
Convective	–	$Pe = \dfrac{UL}{D}$	$St = \dfrac{k_w}{U}$	$Da_I = \dfrac{Lr_A}{U\Delta C}$
Diffusive		–	$Sh = \dfrac{k_w L}{D}$	$Da_{II} = \dfrac{L^2 r_A}{D\Delta C}$
Interface			–	$\dfrac{r_A L}{k_w \Delta C}$
Reaction				–

Table B.3. Dimensionless numbers based on heat flux

	Convective	Conductive	Interphase	Reaction
Heat flux (J s^{-1})	$\rho C_p \Delta T U L^2$	$k \Delta T L$	$h \Delta T L^2$	$r_A L^3 \Delta H$
Convective	–	$Pe = \dfrac{\rho c_p U L}{k}$	$St = \dfrac{h}{c_p \rho U}$	$Da_{III} = \dfrac{r_A L \Delta H}{\rho c_p U \Delta T}$
Conductive		–	$Nu = \dfrac{hL}{k}$	$Da_{IV} = \dfrac{r_A \Delta H L^2}{k \Delta T}$
Interphase			–	$\dfrac{r_A L \Delta H}{h \Delta T}$
Reaction				–

Table B.4. Common dimensionless numbers

Name	Definition	
Biot	$Bi = \dfrac{hL}{k}$	$\dfrac{\text{external heat transfer}}{\text{internal conductive heat transfer}}$
Capillary	$Ca = \dfrac{\mu U}{\gamma}$	$\dfrac{\text{viscous force}}{\text{capillary force}}$
Damköhler I	$Da_I = \dfrac{k_n C_0^n}{C_0 U/L} = k_n C_0^{n-1} \tau$	$\dfrac{\text{reaction rate}}{\text{convective mass transport rate}}$
Damköhler II	$Da_{II} = \dfrac{k_n C_0^{n-1}}{k_g a}$	$\dfrac{\text{reaction rate}}{\text{mass transport rate}}$
Damköhler III	$Da_{III} = \dfrac{k_n C_0^{n-1} L \Delta H}{\rho c_p U \Delta T}$	$\dfrac{\text{heat of reaction}}{\text{heat transport rate}}$
Eotvos	$Eo = \dfrac{gL^2 \Delta \rho}{\gamma}$	$\dfrac{\text{buoyancy force}}{\text{surface tension force}}$
Euler	$Eu = \dfrac{\Delta p}{\rho U^2}$	$\dfrac{\text{pressure force}}{\text{inertial force}}$
Fourier	$Fo = \dfrac{\alpha t}{L^2}$	$\dfrac{\text{heat conduction}}{\text{heat accumulation}}$
Froude	$Fr = \dfrac{gL \Delta \rho}{\rho U^2}$	$\dfrac{\text{buoyed weight}}{\text{surface inertial force}}$
Grashof	$Gr = \dfrac{\rho^2 g \beta (T_s - T_\infty) L^3}{\mu^2}$	$\dfrac{\text{(buoyancy) (inertial force)}}{\text{viscous force}}$
Nusselt	$Nu = \dfrac{hL}{k}$	$\dfrac{\text{convective heat transfer}}{\text{conductive heat transfer}}$
Péclet	$Pe_H = \dfrac{LU}{\alpha}$ or $Pe_D = \dfrac{LU}{D}$	$\dfrac{\text{convective transport}}{\text{diffusive transport}}$
Prandtl	$Pr = \dfrac{\mu}{\rho \alpha}$	$\dfrac{\text{viscosity}}{\text{heat diffusivity}}$
Rayleigh	$Ra = \dfrac{\rho g \beta (T_s - T_\infty) L^3}{\mu \alpha}$	$\dfrac{\text{free convection}}{\text{conduction}}$

(cont.)

Table B.4. (*cont.*)

Name	Definition	
Reynolds	$Re = \dfrac{\rho U L}{\mu}$	$\dfrac{\text{inertial force}}{\text{viscous force}}$
Schmidt	$Sc = \dfrac{\mu}{\rho D}$	$\dfrac{\text{viscosity}}{\text{diffusivity}}$
Sherwood	$Sh = \dfrac{k_c L}{D}$	$\dfrac{\text{convective mass transfer}}{\text{conductive mass transfer}}$
Stokes	$St = \dfrac{\tau_p}{\tau_{fluid}}$	$\dfrac{\text{particle relaxation time}}{\text{fluid characteristic time}}$
Weber	$We = \dfrac{\rho U^2 L}{\gamma}$	$\dfrac{\text{inertial force}}{\text{surface tension force}}$

Appendix C
Student's t-distribution

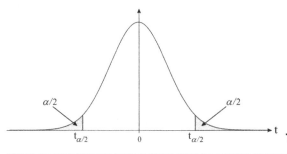

Table C.1. t-table

Degrees of freedom	Probability (Area in one tail, $\frac{\alpha}{2}$)						
	$t_{0.100}$	$t_{0.05}$	$t_{0.025}$	$t_{0.0100}$	$t_{0.0050}$	$t_{0.001}$	$t_{0.0005}$
1	3.08	6.31	12.71	31.82	63.66	318.31	636.62
2	1.89	2.92	4.30	6.96	9.92	22.33	31.60
3	1.64	2.35	3.18	4.54	5.84	10.21	12.92
4	1.53	2.13	2.78	3.75	4.60	7.17	8.61
5	1.48	2.02	2.57	3.36	4.03	5.89	6.87
6	1.44	1.94	2.45	3.14	3.71	5.21	5.96
7	1.41	1.89	2.36	3.00	3.50	4.79	5.41
8	1.40	1.86	2.31	2.90	3.36	4.50	5.04
9	1.38	1.83	2.26	2.82	3.25	4.30	4.78
10	1.37	1.81	2.23	2.76	3.17	4.14	4.59
11	1.36	1.80	2.20	2.72	3.11	4.02	4.44
12	1.36	1.78	2.18	2.68	3.05	3.93	4.32
13	1.35	1.77	2.16	2.65	3.01	3.85	4.22
14	1.34	1.76	2.14	2.62	2.98	3.79	4.14
15	1.34	1.75	2.13	2.60	2.95	3.73	4.07
16	1.34	1.75	2.12	2.58	2.92	3.69	4.01
17	1.33	1.74	2.11	2.57	2.90	3.65	3.97
18	1.33	1.73	2.10	2.55	2.88	3.61	3.92
19	1.33	1.73	2.09	2.54	2.86	3.58	3.88
20	1.33	1.72	2.09	2.53	2.85	3.55	3.85
21	1.32	1.72	2.08	2.52	2.83	3.53	3.82
22	1.32	1.72	2.07	2.51	2.82	3.50	3.79
23	1.32	1.71	2.07	2.50	2.81	3.48	3.77
24	1.32	1.71	2.06	2.49	2.80	3.47	3.75
25	1.32	1.71	2.06	2.49	2.79	3.45	3.73
26	1.32	1.71	2.06	2.48	2.78	3.44	3.71
27	1.31	1.70	2.05	2.47	2.77	3.42	3.69
28	1.31	1.70	2.05	2.47	2.76	3.41	3.67
29	1.31	1.70	2.05	2.46	2.76	3.40	3.66
30	1.31	1.70	2.04	2.46	2.75	3.39	3.65
40	1.30	1.68	2.02	2.42	2.70	3.31	3.55
50	1.30	1.68	2.01	2.40	2.68	3.26	3.50
70	1.29	1.67	1.99	2.38	2.65	3.21	3.44
90	1.29	1.66	1.99	2.37	2.63	3.18	3.40
100	1.29	1.66	1.98	2.36	2.63	3.17	3.39
150	1.29	1.66	1.98	2.35	2.61	3.15	3.36
∞	1.28	1.64	1.96	2.33	2.58	3.09	3.29

C.1   The *F*-distribution

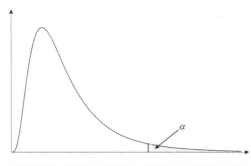

Table C.2. *F*-table; 10% confidence limit

v_1 Degrees of v_2 freedom	1	2	3	4	5	6	7	8	9	10
1	39.86	49.50	53.59	55.83	57.24	58.20	58.91	59.44	59.86	60.20
2	8.53	9.00	9.16	9.24	9.29	9.33	9.35	9.37	9.38	9.39
3	5.54	5.46	5.39	5.34	5.31	5.28	5.27	5.25	5.24	5.23
4	4.54	4.32	4.19	4.11	4.05	4.01	3.98	3.95	3.94	3.92
5	4.06	3.78	3.62	3.52	3.45	3.40	3.37	3.34	3.32	3.30
6	3.78	3.46	3.29	3.18	3.11	3.05	3.01	2.98	2.96	2.94
7	3.59	3.26	3.07	2.96	2.88	2.83	2.78	2.75	2.72	2.70
8	3.46	3.11	2.92	2.81	2.73	2.67	2.62	2.59	2.56	2.54
9	3.36	3.01	2.81	2.69	2.61	2.55	2.51	2.47	2.44	2.42
10	3.28	2.92	2.73	2.61	2.52	2.46	2.41	2.38	2.35	2.32
11	3.23	2.86	2.66	2.54	2.45	2.39	2.34	2.30	2.27	2.25
12	3.18	2.81	2.61	2.48	2.39	2.33	2.28	2.24	2.21	2.19
13	3.14	2.76	2.56	2.43	2.35	2.28	2.23	2.20	2.16	2.14
14	3.10	2.73	2.52	2.39	2.31	2.24	2.19	2.15	2.12	2.10
15	3.07	2.70	2.49	2.36	2.27	2.21	2.16	2.12	2.09	2.06
16	3.05	2.67	2.46	2.33	2.24	2.18	2.13	2.09	2.06	2.03
17	3.03	2.64	2.44	2.31	2.22	2.15	2.10	2.06	2.03	2.00
18	3.01	2.62	2.42	2.29	2.20	2.13	2.08	2.04	2.00	1.98
19	2.99	2.61	2.40	2.27	2.18	2.11	2.06	2.02	1.98	1.96
20	2.97	2.59	2.38	2.25	2.16	2.09	2.04	2.00	1.96	1.94
21	2.96	2.57	2.36	2.23	2.14	2.08	2.02	1.98	1.95	1.92
22	2.95	2.56	2.35	2.22	2.13	2.06	2.01	1.97	1.93	1.90
23	2.94	2.55	2.34	2.21	2.11	2.05	1.99	1.95	1.92	1.89
24	2.93	2.54	2.33	2.19	2.10	2.04	1.98	1.94	1.91	1.88
25	2.92	2.53	2.32	2.18	2.09	2.02	1.97	1.93	1.89	1.87
26	2.91	2.52	2.31	2.17	2.08	2.01	1.96	1.92	1.88	1.86
27	2.90	2.51	2.30	2.17	2.07	2.00	1.95	1.91	1.87	1.85
28	2.89	2.50	2.29	2.16	2.06	2.00	1.94	1.90	1.87	1.84
29	2.89	2.50	2.28	2.15	2.06	1.99	1.93	1.89	1.86	1.83
30	2.88	2.49	2.28	2.14	2.05	1.98	1.93	1.88	1.85	1.82
40	2.84	2.44	2.23	2.09	2.00	1.93	1.87	1.83	1.79	1.76
60	2.79	2.39	2.18	2.04	1.95	1.87	1.82	1.77	1.74	1.71
80	2.77	2.37	2.15	2.02	1.92	1.85	1.79	1.75	1.71	1.68
100	2.76	2.36	2.14	2.00	1.91	1.83	1.78	1.73	1.69	1.66
150	2.74	2.34	2.12	1.98	1.89	1.81	1.76	1.71	1.67	1.64
∞	2.71	2.30	2.08	1.94	1.85	1.77	1.72	1.67	1.63	1.60

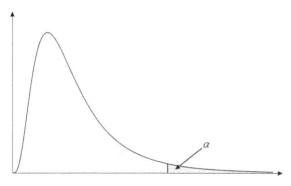

12	15	20	25	30	40	60	100	150	∞
60.71	61.22	61.74	62.05	62.26	62.53	62.79	63.01	63.11	63.33
9.41	9.42	9.44	9.45	9.46	9.47	9.47	9.48	9.48	9.49
5.22	5.20	5.18	5.17	5.17	5.16	5.15	5.14	5.14	5.13
3.90	3.87	3.84	3.83	3.82	3.80	3.79	3.78	3.77	3.76
3.27	3.24	3.21	3.19	3.17	3.16	3.14	3.13	3.12	3.10
2.90	2.87	2.84	2.81	2.80	2.78	2.76	2.75	2.74	2.72
2.67	2.63	2.59	2.57	2.56	2.54	2.51	2.50	2.49	2.47
2.50	2.46	2.42	2.40	2.38	2.36	2.34	2.32	2.31	2.29
2.38	2.34	2.30	2.27	2.25	2.23	2.21	2.19	2.18	2.16
2.28	2.24	2.20	2.17	2.16	2.13	2.11	2.09	2.08	2.06
2.21	2.17	2.12	2.10	2.08	2.05	2.03	2.00	1.99	1.97
2.15	2.10	2.06	2.03	2.01	1.99	1.96	1.94	1.93	1.90
2.10	2.05	2.01	1.98	1.96	1.93	1.90	1.88	1.87	1.85
2.05	2.01	1.96	1.93	1.91	1.89	1.86	1.83	1.82	1.80
2.02	1.97	1.92	1.89	1.87	1.85	1.82	1.79	1.78	1.76
1.99	1.94	1.89	1.86	1.84	1.81	1.78	1.76	1.74	1.72
1.96	1.91	1.86	1.83	1.81	1.78	1.75	1.73	1.71	1.69
1.93	1.89	1.84	1.80	1.78	1.75	1.72	1.70	1.68	1.66
1.91	1.86	1.81	1.78	1.76	1.73	1.70	1.67	1.66	1.63
1.89	1.84	1.79	1.76	1.74	1.71	1.68	1.65	1.64	1.61
1.88	1.83	1.78	1.74	1.72	1.69	1.66	1.63	1.62	1.59
1.86	1.81	1.76	1.73	1.70	1.67	1.64	1.61	1.60	1.57
1.84	1.80	1.74	1.71	1.69	1.66	1.62	1.59	1.58	1.55
1.83	1.78	1.73	1.70	1.67	1.64	1.61	1.58	1.56	1.53
1.82	1.77	1.72	1.68	1.66	1.63	1.59	1.56	1.55	1.52
1.81	1.76	1.71	1.67	1.65	1.61	1.58	1.55	1.54	1.50
1.80	1.75	1.70	1.66	1.64	1.60	1.57	1.54	1.52	1.49
1.79	1.74	1.69	1.65	1.63	1.59	1.56	1.53	1.51	1.48
1.78	1.73	1.68	1.64	1.62	1.58	1.55	1.52	1.50	1.47
1.77	1.72	1.67	1.63	1.61	1.57	1.54	1.51	1.49	1.46
1.71	1.66	1.61	1.57	1.54	1.51	1.47	1.43	1.42	1.38
1.66	1.60	1.54	1.50	1.48	1.44	1.40	1.36	1.34	1.29
1.63	1.57	1.51	1.47	1.44	1.40	1.36	1.32	1.30	1.24
1.61	1.56	1.49	1.45	1.42	1.38	1.34	1.29	1.27	1.21
1.59	1.53	1.47	1.43	1.40	1.35	1.30	1.26	1.23	1.17
1.55	1.49	1.42	1.38	1.34	1.30	1.24	1.18	1.15	1.00

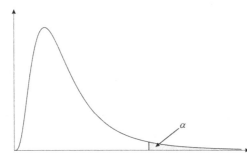

Table C.3. *F*-table; 5% confidence limit

v_1 Degrees of v_2 freedom	1	2	3	4	5	6	7	8	9	10
1	161.45	199.50	215.71	224.58	230.16	233.99	236.77	238.88	240.54	241.88
2	18.51	19.00	19.16	19.25	19.30	19.33	19.35	19.37	19.38	19.40
3	10.13	9.55	9.28	9.12	9.01	8.94	8.89	8.85	8.81	8.79
4	7.71	6.94	6.59	6.39	6.26	6.16	6.09	6.04	6.00	5.96
5	6.61	5.79	5.41	5.19	5.05	4.95	4.88	4.82	4.77	4.74
6	5.99	5.14	4.76	4.53	4.39	4.28	4.21	4.15	4.10	4.06
7	5.59	4.74	4.35	4.12	3.97	3.87	3.79	3.73	3.68	3.64
8	5.32	4.46	4.07	3.84	3.69	3.58	3.50	3.44	3.39	3.35
9	5.12	4.26	3.86	3.63	3.48	3.37	3.29	3.23	3.18	3.14
10	4.96	4.10	3.71	3.48	3.33	3.22	3.14	3.07	3.02	2.98
11	4.84	3.98	3.59	3.36	3.20	3.09	3.01	2.95	2.90	2.85
12	4.75	3.89	3.49	3.26	3.11	3.00	2.91	2.85	2.80	2.75
13	4.67	3.81	3.41	3.18	3.03	2.92	2.83	2.77	2.71	2.67
14	4.60	3.74	3.34	3.11	2.96	2.85	2.76	2.70	2.65	2.60
15	4.54	3.68	3.29	3.06	2.90	2.79	2.71	2.64	2.59	2.54
16	4.49	3.63	3.24	3.01	2.85	2.74	2.66	2.59	2.54	2.49
17	4.45	3.59	3.20	2.96	2.81	2.70	2.61	2.55	2.49	2.45
18	4.41	3.55	3.16	2.93	2.77	2.66	2.58	2.51	2.46	2.41
19	4.38	3.52	3.13	2.90	2.74	2.63	2.54	2.48	2.42	2.38
20	4.35	3.49	3.10	2.87	2.71	2.60	2.51	2.45	2.39	2.35
21	4.32	3.47	3.07	2.84	2.68	2.57	2.49	2.42	2.37	2.32
22	4.30	3.44	3.05	2.82	2.66	2.55	2.46	2.40	2.34	2.30
23	4.28	3.42	3.03	2.80	2.64	2.53	2.44	2.37	2.32	2.27
24	4.26	3.40	3.01	2.78	2.62	2.51	2.42	2.36	2.30	2.25
25	4.24	3.39	2.99	2.76	2.60	2.49	2.40	2.34	2.28	2.24
26	4.23	3.37	2.98	2.74	2.59	2.47	2.39	2.32	2.27	2.22
27	4.21	3.35	2.96	2.73	2.57	2.46	2.37	2.31	2.25	2.20
28	4.20	3.34	2.95	2.71	2.56	2.45	2.36	2.29	2.24	2.19
29	4.18	3.33	2.93	2.70	2.55	2.43	2.35	2.28	2.22	2.18
30	4.17	3.32	2.92	2.69	2.53	2.42	2.33	2.27	2.21	2.16
40	4.08	3.23	2.84	2.61	2.45	2.34	2.25	2.18	2.12	2.08
60	4.00	3.15	2.76	2.53	2.37	2.25	2.17	2.10	2.04	1.99
80	3.96	3.11	2.72	2.49	2.33	2.21	2.13	2.06	2.00	1.95
100	3.94	3.09	2.70	2.46	2.31	2.19	2.10	2.03	1.97	1.93
150	3.90	3.06	2.66	2.43	2.27	2.16	2.07	2.00	1.94	1.89
∞	3.84	3.00	2.60	2.37	2.21	2.10	2.01	1.94	1.88	1.83

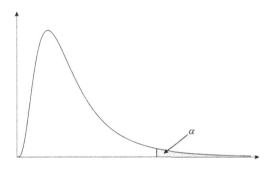

12	15	20	25	30	40	60	100	150	∞
243.91	245.95	248.01	249.26	250.10	251.14	252.20	253.04	253.46	254.31
19.41	19.43	19.45	19.46	19.46	19.47	19.48	19.49	19.49	19.50
8.74	8.70	8.66	8.63	8.62	8.59	8.57	8.55	8.54	8.53
5.91	5.86	5.80	5.77	5.75	5.72	5.69	5.66	5.65	5.63
4.68	4.62	4.56	4.52	4.50	4.46	4.43	4.41	4.39	4.36
4.00	3.94	3.87	3.83	3.81	3.77	3.74	3.71	3.70	3.67
3.57	3.51	3.44	3.40	3.38	3.34	3.30	3.27	3.26	3.23
3.28	3.22	3.15	3.11	3.08	3.04	3.01	2.97	2.96	2.93
3.07	3.01	2.94	2.89	2.86	2.83	2.79	2.76	2.74	2.71
2.91	2.84	2.77	2.73	2.70	2.66	2.62	2.59	2.57	2.54
2.79	2.72	2.65	2.60	2.57	2.53	2.49	2.46	2.44	2.40
2.69	2.62	2.54	2.50	2.47	2.43	2.38	2.35	2.33	2.30
2.60	2.53	2.46	2.41	2.38	2.34	2.30	2.26	2.24	2.21
2.53	2.46	2.39	2.34	2.31	2.27	2.22	2.19	2.17	2.13
2.48	2.40	2.33	2.28	2.25	2.20	2.16	2.12	2.10	2.07
2.42	2.35	2.28	2.23	2.19	2.15	2.11	2.07	2.05	2.01
2.38	2.31	2.23	2.18	2.15	2.10	2.06	2.02	2.00	1.96
2.34	2.27	2.19	2.14	2.11	2.06	2.02	1.98	1.96	1.92
2.31	2.23	2.16	2.11	2.07	2.03	1.98	1.94	1.92	1.88
2.28	2.20	2.12	2.07	2.04	1.99	1.95	1.91	1.89	1.84
2.25	2.18	2.10	2.05	2.01	1.96	1.92	1.88	1.86	1.81
2.23	2.15	2.07	2.02	1.98	1.94	1.89	1.85	1.83	1.78
2.20	2.13	2.05	2.00	1.96	1.91	1.86	1.82	1.80	1.76
2.18	2.11	2.03	1.98	1.94	1.89	1.84	1.80	1.78	1.73
2.16	2.09	2.01	1.96	1.92	1.87	1.82	1.78	1.76	1.71
2.15	2.07	1.99	1.94	1.90	1.85	1.80	1.76	1.74	1.69
2.13	2.06	1.97	1.92	1.88	1.84	1.79	1.74	1.72	1.67
2.12	2.04	1.96	1.91	1.87	1.82	1.77	1.73	1.70	1.65
2.10	2.03	1.94	1.89	1.85	1.81	1.75	1.71	1.69	1.64
2.09	2.01	1.93	1.88	1.84	1.79	1.74	1.70	1.67	1.62
2.00	1.92	1.84	1.78	1.74	1.69	1.64	1.59	1.56	1.51
1.92	1.84	1.75	1.69	1.65	1.59	1.53	1.48	1.45	1.39
1.88	1.79	1.70	1.64	1.60	1.54	1.48	1.43	1.39	1.32
1.85	1.77	1.68	1.62	1.57	1.52	1.45	1.39	1.36	1.28
1.82	1.73	1.64	1.58	1.54	1.48	1.41	1.34	1.31	1.22
1.75	1.67	1.57	1.51	1.46	1.39	1.32	1.24	1.20	1.00

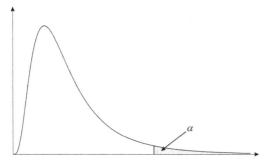

Table C.4. *F*-table; 1% confidence limit

v_1 Degrees of freedom v_2	1	2	3	4	5	6	7	8	9	10
1	4 052	5 000	5 403	5 625	5 764	5 859	5 928	5 981	6 022	6 056
2	98.50	99.00	99.17	99.25	99.30	99.33	99.36	99.37	99.39	99.40
3	34.12	30.82	29.46	28.71	28.24	27.91	27.67	27.49	27.35	27.23
4	21.20	18.00	16.69	15.98	15.52	15.21	14.98	14.80	14.66	14.55
5	16.26	13.27	12.06	11.39	10.97	10.67	10.46	10.29	10.16	10.05
6	13.74	10.92	9.78	9.15	8.75	8.47	8.26	8.10	7.98	7.87
7	12.25	9.55	8.45	7.85	7.46	7.19	6.99	6.84	6.72	6.62
8	11.26	8.65	7.59	7.01	6.63	6.37	6.18	6.03	5.91	5.81
9	10.56	8.02	6.99	6.42	6.06	5.80	5.61	5.47	5.35	5.26
10	10.04	7.56	6.55	5.99	5.64	5.39	5.20	5.06	4.94	4.85
11	9.65	7.21	6.22	5.67	5.32	5.07	4.89	4.74	4.63	4.54
12	9.33	6.93	5.95	5.41	5.06	4.82	4.64	4.50	4.39	4.30
13	9.07	6.70	5.74	5.21	4.86	4.62	4.44	4.30	4.19	4.10
14	8.86	6.51	5.56	5.04	4.70	4.46	4.28	4.14	4.03	3.94
15	8.68	6.36	5.42	4.89	4.56	4.32	4.14	4.00	3.89	3.80
16	8.53	6.23	5.29	4.77	4.44	4.20	4.03	3.89	3.78	3.69
17	8.40	6.11	5.18	4.67	4.34	4.10	3.93	3.79	3.68	3.59
18	8.29	6.01	5.09	4.58	4.25	4.01	3.84	3.71	3.60	3.51
19	8.18	5.93	5.01	4.50	4.17	3.94	3.77	3.63	3.52	3.43
20	8.10	5.85	4.94	4.43	4.10	3.87	3.70	3.56	3.46	3.37
21	8.02	5.78	4.87	4.37	4.04	3.81	3.64	3.51	3.40	3.31
22	7.95	5.72	4.82	4.31	3.99	3.76	3.59	3.45	3.35	3.26
23	7.88	5.66	4.76	4.26	3.94	3.71	3.54	3.41	3.30	3.21
24	7.82	5.61	4.72	4.22	3.90	3.67	3.50	3.36	3.26	3.17
25	7.77	5.57	4.68	4.18	3.86	3.63	3.46	3.32	3.22	3.13
26	7.72	5.53	4.64	4.14	3.82	3.59	3.42	3.29	3.18	3.09
27	7.68	5.49	4.60	4.11	3.78	3.56	3.39	3.26	3.15	3.06
28	7.64	5.45	4.57	4.07	3.75	3.53	3.36	3.23	3.12	3.03
29	7.60	5.42	4.54	4.04	3.73	3.50	3.33	3.20	3.09	3.00
30	7.56	5.39	4.51	4.02	3.70	3.47	3.30	3.17	3.07	2.98
40	7.31	5.18	4.31	3.83	3.51	3.29	3.12	2.99	2.89	2.80
60	7.08	4.98	4.13	3.65	3.34	3.12	2.95	2.82	2.72	2.63
80	6.96	4.88	4.04	3.56	3.26	3.04	2.87	2.74	2.64	2.55
100	6.90	4.82	3.98	3.51	3.21	2.99	2.82	2.69	2.59	2.50
150	6.81	4.75	3.91	3.45	3.14	2.92	2.76	2.63	2.53	2.44
∞	6.63	4.61	3.78	3.32	3.02	2.80	2.64	2.51	2.41	2.32

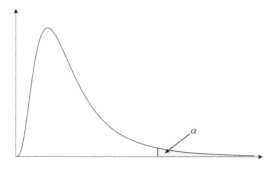

12	15	20	25	30	40	60	100	150	∞
6.106	6.157	6.209	6.240	6.261	6.287	6.313	6.334	6.345	6.366
99.42	99.43	99.45	99.46	99.47	99.47	99.48	99.49	99.49	99.50
27.05	26.87	26.69	26.58	26.50	26.41	26.32	26.24	26.20	26.13
14.37	14.20	14.02	13.91	13.84	13.75	13.65	13.58	13.54	13.46
9.89	9.72	9.55	9.45	9.38	9.29	9.20	9.13	9.09	9.02
7.72	7.56	7.40	7.30	7.23	7.14	7.06	6.99	6.95	6.88
6.47	6.31	6.16	6.06	5.99	5.91	5.82	5.75	5.72	5.65
5.67	5.52	5.36	5.26	5.20	5.12	5.03	4.96	4.93	4.86
5.11	4.96	4.81	4.71	4.65	4.57	4.48	4.42	4.38	4.31
4.71	4.56	4.41	4.31	4.25	4.17	4.08	4.01	3.98	3.91
4.40	4.25	4.10	4.01	3.94	3.86	3.78	3.71	3.67	3.60
4.16	4.01	3.86	3.76	3.70	3.62	3.54	3.47	3.43	3.36
3.96	3.82	3.66	3.57	3.51	3.43	3.34	3.27	3.24	3.17
3.80	3.66	3.51	3.41	3.35	3.27	3.18	3.11	3.08	3.00
3.67	3.52	3.37	3.28	3.21	3.13	3.05	2.98	2.94	2.87
3.55	3.41	3.26	3.16	3.10	3.02	2.93	2.86	2.83	2.75
3.46	3.31	3.16	3.07	3.00	2.92	2.83	2.76	2.73	2.65
3.37	3.23	3.08	2.98	2.92	2.84	2.75	2.68	2.64	2.57
3.30	3.15	3.00	2.91	2.84	2.76	2.67	2.60	2.57	2.49
3.23	3.09	2.94	2.84	2.78	2.69	2.61	2.54	2.50	2.42
3.17	3.03	2.88	2.78	2.72	2.64	2.55	2.48	2.44	2.36
3.12	2.98	2.83	2.73	2.67	2.58	2.50	2.42	2.38	2.31
3.07	2.93	2.78	2.69	2.62	2.54	2.45	2.37	2.34	2.26
3.03	2.89	2.74	2.64	2.58	2.49	2.40	2.33	2.29	2.21
2.99	2.85	2.70	2.60	2.54	2.45	2.36	2.29	2.25	2.17
2.96	2.82	2.66	2.57	2.50	2.42	2.33	2.25	2.21	2.13
2.93	2.78	2.63	2.54	2.47	2.38	2.29	2.22	2.18	2.10
2.90	2.75	2.60	2.51	2.44	2.35	2.26	2.19	2.15	2.06
2.87	2.73	2.57	2.48	2.41	2.33	2.23	2.16	2.12	2.03
2.84	2.70	2.55	2.45	2.39	2.30	2.21	2.13	2.09	2.01
2.66	2.52	2.37	2.27	2.20	2.11	2.02	1.94	1.90	1.80
2.50	2.35	2.20	2.10	2.03	1.94	1.84	1.75	1.70	1.60
2.42	2.27	2.12	2.01	1.94	1.85	1.75	1.65	1.61	1.49
2.37	2.22	2.07	1.97	1.89	1.80	1.69	1.60	1.55	1.43
2.31	2.16	2.00	1.90	1.83	1.73	1.62	1.52	1.46	1.33
2.18	2.04	1.88	1.77	1.70	1.59	1.47	1.36	1.29	1.00

Bibliography

Bird, R.B., W.E. Stewart, and E.N. Lightfoot, *Transport Phenomena*, 2nd edn. New York: Wiley, 2002.

Levenspiel, O., *Chemical Reaction Engineering*, 3rd edn. New York: Wiley, 1999.

Welty, J.R., C.E. Wicks, and R.E. Wilson, *Fundamentals of Momentum, Heat and Mass Transfer*, 5th edn. Hoboken, NJ: Wiley, 2007.

Kunes, J., *Dimensionless Physical Quantities in Science and Engineering*. Burlington, MA: Elsevier, 2012.

LeVeque, R.J., *Finite Difference Methods for Ordinary and Partial Differential Equations: Steady-State and Time-Dependent Problems*. Philadelphia, PA: SIAM, 2007.

Sauer, T., *Numerical Analysis*. Boston, MA: Pearson, 2012.

Haberman, R., *Applied Partial Differential Equations*. Upper Saddle River, NJ: Pearson, 2004.

Draper, N.R. and H. Smith, *Applied Regression Analysis*, 3rd edn. New York: John Wiley & Sons, 1998.

Chatfield, C., *Statistics for Technology*, 3rd edn. London: Chapman and Hall, 1983.

Mendenhall, W. and T. Sincich, *A Second Course in Statistics Regression Analysis*, 6th edn. Upper Saddle River, NJ: Pearson Education International, 2003.

Devore, J. and N. Farnum, *Applied Statistics for Engineers and Scientists*, 2nd edn. Belmont, CA: Thomson, 2005.

Montgomery, D.C., *Design and Analysis of Experiments*, 7th edn. Hoboken, NJ: John Wiley & Sons, 2009.

Index

accumulation, 23
accuracy, 88
Adams–Bashforth method, 93
Adams–Moulton method, 94, 97
adaptive step size, 88
ANOVA table, 145, 146, 152, 156
approximate confidence interval, 140
averaging, 54

backward difference, 113
backward differentiation formulas, 97
balance, 20
    component, 21
    energy, 21
    mass, 20
    momentum, 22
base dimensions, 40
boundary conditions, 26, 54, 67
boundary-value problem, 82, 99
Buckingham's Π theorem, 48

central differences, 103, 110
characteristic variables, 44
circulation time, 51
collocation, 107
collocation points, 107
conditional stability, 90
conduction, 23
confidence band, 135, 136, 141, 161
confidence interval, 130, 152, 156, 162
    for the dependent variable, 135
confidence region, 135, 141, 157, 162
conservation, 20
constitutive equation, 22
    Darcy, 22
    Fick, 22
    Fourier, 22
    Newton, 22
continuous variable, 11
control volume, 24
correlation, 40, 136, 142, 153, 158, 162
correlation matrix, 136, 137, 142
coupling, 14
Crank–Nicolson method, 114

decoupling, 54, 55, 72
degrees of freedom, 130, 131
dependent variable, 121
deterministic model, 13
diffusion, 24
dimensionless equations, 41, 43
dimensionless number
    Euler, 43
    Froude, 43
    Péclet, 43
    Prandtl, 43
    Reynolds, 43
    Schmidt, 43
dimensionless variable, 40, 46
Dirichlet boundary condition, 103
discrete variable, 11
distributed parameter, 11
Dormand–Prince method, 89

elliptic PDE, 110
empirical model, 14, 40, 46
error, 53
    global, 87
    local, 87
error control, 89
error estimation, 76
error tolerance, 98
explicit algorithms, 87
explicit Euler method, 82
extrapolation, 13

F-distribution, 135
fictitious boundary, 104, 106
film theory, 45
finite difference method, 102, 110
finite element method, 107
first-order derivative, 64
flow, 24
flux, 22
forward difference, 110
Frössling correlation, 46

Gaussian elimination, 103, 106
grid, 102

Hagen–Poiseuille equation, 44
heat transfer, 44, 46
higher-order differential equations, 96
hydraulic diameter, 57
hyperbolic PDE, 110

implicit Euler method, 91
implicit methods, 90
independent variable, 54, 69, 121
initial-value problem, 82
interpolation, 13
intrinsically linear models, 138

Jacobian, 140

lack of fit, 146–148, 152
least square method, 123, 126
limiting cases, 54, 63
linear model, 10
linear regression, 121, 122, 125
linearizing, 54, 61
local truncation error, 83, 87
lumped parameter, 11
lumping, 56

MATLAB, 114
mechanistic model, 14
mesh, 102
micromixing, 51
midpoint method, 84
mirror planes, 56
mixing, 50
multistep methods
    explicit, 93
    implicit, 94

neglecting terms, 54, 64
Neumann boundary condition, 104
Newton method, 93
non-linear model, 10, 123, 139, 159
non-recurring set, 47
normal equation, 124
numerical differentiation formulas, 97

Octave, 117
ordinary differential equation, 81
outlier, 129, 145
over- and underestimation, 77

parabolic PDE, 110
partial differential equations, 108
periodic boundary condition, 56
polynomial model, 122
population balance, 28
predictor, 121
predictor–corrector pairs, 94

primary dimensions, 46
pumping capacity, 51
pure error, 146, 147

R^2 statistic, 149
random error, 132
Ranz–Marshall correlation, 46
rate determining, 50
rate of dissipation, 51
reaction time, 50
recurring set, 47
regressor, 121
residence time distribution, 32
residual plots, 142–144
response variable, 121
Runge–Kutta method, 86
Runge–Kutta–Fehlberg method, 89

scale
    macroscopic, 18
    mesoscopic, 17
    microscopic, 16
    molecular, 16
scale-independent model, 49
scaling down, 48
scaling out, 48
scaling rule, 51
scaling up, 48
second-order derivative, 64
sensitivity analysis, 77
shell balance, 23
shooting method, 99
simplified geometry, 54
stability, 90, 91
standard error, 130
steady state, 11, 54, 58, 60, 70
stencil, 111
stiff problems, 97
stochastic model, 13
straight-line model, 121, 123, 131
Student's t-distribution, 133
Studentized residual, 150
sum of square error, 123, 124, 126, 130, 147
symmetry, 55, 56, 69
system of differential equations, 94

Taylor expansion, 62, 83
t-distribution, 135
time constants, 58
time scales, 97
total corrected sum of squares, 148
transformation, 96
transient model, 11, 54, 58
transport phenomena, 22
trial function, 107
truncation error, 87

t-tests of individual parameters, 133
turbulent eddies, 51
turbulent kinetic energy, 51

uncertainty, 53
unconditional stability, 91
unsteady state, 58

validation, 53
variance, 125, 127, 136
verification, 53

weighted least squares, 127
weighted sum of square error,
     129

Printed in the United States
By Bookmasters